Boukar Michel

The impact of digitalization on rural economies in Africa

Boukar Michel

The impact of digitalization on rural economies in Africa

The role of artificial intelligence in precision agriculture Preface by Mr. Lacina Koné

ScienciaScripts

Cover image: Provided by the author

This book is a translation from the original published under ISBN 978-620-6-73139-9.

Publisher:
Sciencia Scripts
is a trademark of
Dodo Books Indian Ocean Ltd. and OmniScriptum S.R.L publishing group

120 High Road, East Finchley, London, N2 9ED, United Kingdom
Str. Armeneasca 28/1, office 1, Chisinau MD-2012, Republic of Moldova, Europe
Managing Directors: Ieva Konstantinova, Victoria Ursu
info@omniscriptum.com

Printed at: see last page
ISBN: 978-620-8-63895-5

The impact of digitalization on rural economies in Africa

The role of artificial intelligence in precision agriculture

Dr BOUKAR Michel

BY THE SAME AUTHOR

A. EUROPEAN UNIVERSITY PUBLISHING HOUSE

1. Michel, Boukar, and Roland Tete. Biofuel production from balanites. European University Publishing, 2019.

2. Michel, Boukar. Agriculture et Énergie Renouvelable post Covid-19: Solutions Pratiques et Perspective pour la Sécurité Alimentaire; Adaptation aux changements climatiques Préface de Djimé Adoum, Ph. D, SE du CILSS. European University Publishing, 2020.

3. Boukar, Michel, Modélisation des émissions de CO_2 dues à la consommation des produits pétroliers dans le transport routier tchadien, Preface by Monsieur Tamba Jean Gaston, Éditions universitaires européennes (20.08.2024)

4. Boukar, Michel, Développement durable au Tchad: es enjeux des financements des partenaires Preface by Prof. Ing. Ében-Ezer César Léonce Koffi, Sept. 2024.

5. Boukar Michel, L'intelligence artificielle et Education Numérique en Afrique : Réduire la fracture numérique par des approches innovantes, Preface by Dr Tom Erdimi, sep 2024

6 Boukar Michel L'impact de la digitalisation sur les économies rurales e-Afrique : Le rôle de l'intelligence artificielle dans l'agriculture de précision, Preface LACINA KONE

Managing Director of SMART AFRICA, European University Publishing, 2025.

B. PUBLISHED BY L'HARMATTAN, PARIS :

1. Boukar, Michel. Waste management in African cities: Reduce, recycle, reuse 3R. Valorisation pour la production d'électricité et autres applications (2023): 1-282. Preface by Professor Maurice Kamto

2. Boukar, Michel. Transition to a green economy. Bâtir une gouvernance pour l'Afrique et le Tchad, 2024.

PREFACE

In an ever-changing world, digitalization is emerging as an unavoidable vector of transformation, affecting all sectors of society, including rural economies. In Africa, this phenomenon is taking on particular significance, offering unprecedented prospects for farming communities that have long been marginalized. Dr BOUKAR Michel, who has demonstrated his talent as a scientific writer in several books, sets out in this book to explore the impact of digitalization on these rural economies, with a particular focus on the crucial role of artificial intelligence in precision agriculture.

Agriculture, always an essential pillar of the African economy, is facing major challenges: climate change, population growth, and pressure on natural resources. In this context, the integration of digital technologies and artificial intelligence appears to be a promising response. Precision farming can optimize yields, reduce waste and improve the sustainability of agricultural practices. Thanks to massive data analysis and innovative tools, farmers can make informed decisions that maximize production while preserving the environment.

This precision agriculture, using advanced technologies such as sensors, drones and data processing tools, enables the collection of specific information on crop conditions. Combined with Artificial Intelligence (AI), it looks like the ultimate solution to the problems facing African farmers.

Indeed, AI plays a crucial role in analyzing this data, enabling farmers to make informed decisions on crop management, anticipate weather conditions and reduce the risks associated with disease or pests. This leads to improved yields and more efficient use of resources.

The effects of digitalization on rural economies are numerous. It boosts agricultural yields, optimizing farmers' profits. Digitization also strengthens value chains by improving market transparency and efficiency. In addition, it can contribute to greater social inclusion, by facilitating access to information and financing, particularly for women and young farmers.

In the depths of its lines, this book is intended as a guide to understanding how these technologies can transform African rural economies. Through case studies and in-depth analysis, Dr BOUKAR Michel highlights the opportunities offered by digitalization and the challenges to be met to ensure a successful transition to smart, sustainable agriculture.

It hopes to stimulate constructive dialogue around the policies needed to support this transformation. With this in mind, it is imperative that decision-makers, investors and civil society players work together to create a favorable ecosystem for technological innovation in the agricultural sector.

We invite you to dive into these pages with an open and curious mind. May this book be a source of inspiration and a call to action for all those who believe in a future where technology and tradition combine to strengthen Africa's rural economies.

LACINA KONE

Managing Director, SMART AFRICA

ABBREVIATIONS

IUT: International Telecommunication Union

UNESCO: United Nations Educational, Scientific and Cultural Organization

FAO: Food and Agriculture Organization of the United Nations

AI: Artificial Intelligence

OECD: Organisation for Economic Co-operation and Development.

USAID: United States Agency for International Development

AU: African Union

EU: European Union

UN: United Nations

AfDB: African Development Bank

IFC: International Finance Corporation

RGPD: General Data Protection Regulation

IEA: International Energy Agency

GSMA: Global System for Mobile Communications

Summary

Digitization is significantly transforming rural economies in Africa, especially in the agricultural sector, where a large proportion of the population depends on farming for their livelihood. The integration of artificial intelligence (AI) into precision agriculture is emerging as a key element of this transformation, offering innovative solutions to improve agricultural yields, optimize resource use and enhance sustainability. The concept of artificial intelligence, AI, is that machines are capable of performing tasks in a presumed intelligent manner. These processes enable them to function automatically, to reason and to learn for themselves. The rise of digital technologies, in particular growing access to the Internet and cell phones, is an important lever for farmers in Africa. These technologies enable them to better access markets, obtain valuable information on prices and improve their decision-making. However, challenges remain, notably in terms of infrastructure, connectivity and education, sometimes limiting the widespread adoption of digital solutions. Precision farming, which involves the use of advanced technologies such as sensors, drones and data processing tools, enables the collection of specific information on crop conditions. AI plays a crucial role in analyzing this data, enabling farmers to make informed decisions on crop management, anticipate climatic conditions and reduce the risks associated with disease or pests. This leads to improved yields and more efficient use of resources. The effects of digitalization on rural economies are numerous. It boosts agricultural yields, optimizing farmers' profits. Digitization also strengthens value chains by improving market transparency and efficiency. In addition, it can contribute to greater social inclusion, by facilitating access to information and financing, particularly for women and young farmers.

However, challenges remain, including the need to train farmers in the use of these new technologies and to invest in the necessary infrastructure to ensure adequate connectivity. In addition, the issue of sustainability and the environmental impacts of the intensive use of technologies must be taken into account. These challenges can be grouped into several key categories: infrastructure and connectivity; education and skills; cost of access and implementation; sustainability and environmental impact; cultural reaction and resistance to change; and the regulatory and policy framework.

One of the main obstacles to implementing digitalization is the lack of basic infrastructure, such as access to the Internet and electricity. In many rural areas, connectivity remains limited, preventing farmers from accessing essential information and digital tools. Without a solid infrastructure, the effectiveness of AI-based agricultural solutions cannot be fully exploited.

Training farmers to use new technologies is a major challenge. Many farmers lack the skills needed to integrate AI and other digital tools into their farming practices. Without an adequate education program and training sessions, adoption of these technologies will be limited. The costs associated with acquiring and implementing digital technologies and AI can be prohibitive for many farmers, especially smallholders. This can create a gap between farmers who are able to invest in these technologies and those who cannot, exacerbating inequalities in the agricultural sector.

The intensive adoption of digital technologies can lead to sustainability concerns. Excessive use of data and resources to collect and analyze information can have an environmental impact, and AI, if misapplied, can lead to agricultural practices that harm biodiversity and local ecosystems. In some rural communities, there may be resistance to adopting new technologies due to deeply rooted traditions or fear of the unknown. Farmers may be skeptical about the effectiveness of AI and digital tools, preferring to stick to traditional farming methods. A lack of political support and appropriate regulatory frameworks can also hinder the adoption of digitization. The absence of innovation-friendly policies can slow the development of the necessary infrastructure and limit incentives for the adoption of advanced technologies.

For digitalization to have a positive impact on rural economies in Africa, it is essential to address these challenges systematically. Investment in infrastructure, farmer training and strong supportive policies are needed to ensure that the benefits of artificial intelligence and precision farming are available to all, particularly smallholders. With this in mind, an integrated and inclusive approach will be fundamental to realizing the potential of digitalization in the agricultural sector.

As for prospects, the impact of digitization on rural economies in Africa, particularly through the integration of artificial intelligence (AI) in precision agriculture, presented important prospects for the future of the agricultural sector and the transformation of local economies. AI makes it possible to optimize farming practices by analyzing precise data on soil, crops and climatic conditions. This ability to provide personalized recommendations could transform traditional farming methods, leading to significant increases in yields. Farmers could thus maximize production while minimizing excessive use of fertilizers and water.

Digitization facilitates farmers' access to markets, thanks to digital platforms that cut out intermediaries and offer real-time price information. This enables them to better position their products and improve profitability. AI-powered early warning systems can also provide weather forecasts and crop management advice. Integrating AI into precision agriculture can strengthen agricultural value chains by improving traccability and transparency. This can help reduce post-harvest losses and ensure better product quality, which is essential for access to international markets. Optimized use of natural resources through technologies such as AI promotes more sustainable agriculture. By enabling more precise management of water and soil resources, this digitization can help preserve the environment while meeting growing food needs.

To maximize the benefits of this digitalization, it is crucial to invest in training programs that enable farmers to familiarize themselves with the new technologies. Promoting inclusion, particularly for women and young people, will be essential to ensure that all farmers can benefit from technological innovations.

Governments and international organizations must play a key role by creating a favorable regulatory framework and investing in the necessary infrastructure, such as Internet access and electricity. This will support the widespread adoption of digital technologies in rural areas.

The prospects offered by digitalization and artificial intelligence in precision agriculture in Africa are promising. They can not only transform rural economies, but also contribute to food security and environmental sustainability. However, for these benefits to materialize, it is imperative to overcome the challenges associated with infrastructure, education and equitable access to technologies.

In short, digitalization and artificial intelligence represent significant opportunities for the development of rural economies in Africa, particularly through precision agriculture. To maximize these benefits, concerted efforts will be needed to address infrastructure, education and sustainability issues to ensure a positive and inclusive transformation of the agricultural sector.

Introduction

[1]A phenomenon of modern times, digitization or "digital transformation" is advancing at lightning speed. It has had a major impact on humankind, bringing about profound changes in our society and, of course, fostering the emergence of new paradigms. It has established itself as an inescapable process that can be applied to all fields, enabling time and money to be optimized by automating increasingly complex tasks. Economic operators in all sectors have made digital transformation a competitive factor, on the one hand to improve their governance, and on the other to be in tune with their customers and the market in general, in real time[2] . The whole world has already begun the transformation towards the digitalization of work tools via software and applications. This transformation has led to changes in the way we live, work and interact. Africa, like other regions of the world, has only recently been hit by this wave of digitalization. The growth of digitalization in Africa is being driven by government initiatives to promote technological innovation, and by the exponential growth in demand from a very young population. As a result, African companies are able to offer more complete, less burdensome and higher quality processes. This growth exposes individuals and organizations to security risks such as data theft, online fraud and cybercriminal attacks.

---

[1] Ahmed Elhanafi, in tribune libre, September 2021; ite.ma

[2] Idem.

Today, 90% of companies in Africa do not use the cyber security protocols necessary to protect their employees. Digitization enables companies to grow thanks to the opportunities it offers.

However, there are major disparities within Africa itself when it comes to digitalization. North Africa and South Africa are already well advanced in their digital transformation, but Central Africa and East Africa are not. Compared to other continents, the divide is y clearly visible. In terms of Internet penetration rates on the continent, only two African countries are above the global average: Morocco and Egypt. Digitization is a transformation process that aims to integrate information and communication technologies (ICT) into the various sectors of society, to improve efficiency, productivity and competitiveness. In Africa, digitization is seen as an essential strategy for promoting economic, social and environmental development.

[3]According to a study by the African Development Bank (AfDB) entitled "Digitization in Africa: a boost for development" (2020), digitization can bring several benefits to Africa, including: digitization can :

- Create new jobs in the technology, e-commerce and service sectors
- To help African companies accelerate their growth by improving their productivity and competitiveness;

[3] African Development Bank (2020). Digitization in Africa: an impetus for development

- Enable African governments to better manage their public affairs, thus improving efficiency and transparency
- Enable African children to have access to quality education, wherever they are in the world;
- Help improve healthcare services in Africa, by enabling patients to consult doctors remotely and providing accurate medical information.

African countries are gradually embracing digitalization. [4]According to a report by the Economic Commission for Africa (ECA) entitled "Digitization in Africa" (2020), 35% of African countries have deployed 4G or 5G telecommunications infrastructures, while 20% have set up national digitization programs. There are many examples of African countries that have made significant progress in digitalization:

- Ethiopia: Ethiop Telecom has launched a national digitization program to improve access to the Internet and financial services;
- Côte d'Ivoire: Côte d'Ivoire has launched an ICT development program to improve access to the Internet and financial services;
- Kenya: Kenya is considered one of the leaders in digitalization in Africa, with a high mobile banking take-up rate.

[4] Economic Commission for Africa (2020). Digitization in Africa

Africa's rural economy is an important part of the continent's economy, with over 60% of the population living in rural areas. However, Africa's rural economy often faces numerous challenges, such as poverty, hunger, disease and fragile production systems. Poverty is endemic in rural Africa, with 35% of the population living on less than US$1.90 a day (World Bank, 2020)[5] ; agricultural productivity is generally low in Africa, due to poor access to arable land, water and technology (FAO, 2020). Many African farmers live in a subsistence situation, meaning they produce enough for their own consumption, but not to sell on markets (World Bank, 2020)[6] . Africa's rural economy is mainly based on agriculture, which accounts for around 20% of GDP in most African countries (FAO, 2020)[7] . Agricultural production is often characterized by small-scale farming, low productivity and high vulnerability to climatic hazards (IFAD, 2020)[8]. African farmers mainly produce cereals, pulses, vegetable oils and dairy products for their own consumption or for the local market. The challenges facing Africa's rural economy are numerous and complex: African farmers are highly vulnerable to climatic hazards such as droughts, floods and storms, which can wipe out their harvests and their well-being (IPCC, 2013)[9] ; African farmers often find it difficult to access markets to sell their produce, due to poor infrastructure, lack of storage capacity and competition with imported products (World

---

[5] World Bank (2020).Africa's Pulse :the impacts of COVID19 in Africa's Economy.

[6] Idem.

[7] FAO (2020). State of Food Security and Nutrition in the world.

[8] IFAD (2020). Rural Poverty Report.

[9] IPCC (2013). Climate Change 2013: The Physical Science Basis.

Bank, 2019)[10] ; African farmers often struggle to access financial services, due to the lack of financial institutions present in rural areas (IFC, 2019)[11] ; ineffective governance can lead to misallocation of resources, which can reduce productivity and economic growth (World Bank, 2019) ; rapid demographic change in Africa is creating pressure on existing natural resources and infrastructure, which can have a negative impact on the rural economy (UNDESA, 2019)[12]Precision agriculture, also known as smart agriculture, is an innovative approach that aims to improve agricultural efficiency and productivity using information and communication technologies (ICT). This approach is particularly important for rural development, as it enables farmers to meet the challenges associated with agricultural production, such as food security, rural poverty and the impact of climate change. [13]Precision farming offers several advantages for rural development, including: precision agriculture enables farmers to better understand their land and crops, which in turn improves their productivity and income; precision agriculture enables farmers to reduce the costs associated with irrigation, pesticides and fertilizers, by using more precise and optimal quantities; precision farming enables farmers to manage their resources (water, fertilizers, pesticides) more efficiently, reducing environmental impact and protecting natural resources; precision

[10] World Bank (2019).

[11] IFC (2019). Africa's Agriculture and Food Security.

[12]UNDESA (2019). World Population Prospects 2019.

[13] Ditto

farming enables farmers to produce more food, improving food security in rural areas; precision farming can create jobs in rural areas, creating employment opportunities in sectors linked to smart farming. Here are just a few examples of the use of precision agriculture in rural development: Ethiopia has set up a precision agriculture program to improve agricultural productivity and food security in rural areas; Kenya has created a smart agriculture program to improve agricultural productivity and reduce irrigation costs; India has set up a precision agriculture program to improve agricultural productivity and reduce agricultural production costs.

Source *: https://agri.vipros.fr/mag/comment-ameliorer-la-gestion-de-leau-avec-un-systeme-dirrigation-intelligent/*

Figure1 : Intelligent irrigation system

[14]Precision farming systems dynamically develop management systems. They enable farmers to solve a whole range of problems. At

[14] Journal of Agricultural science and technology (2019). Precision agriculture: A Review of its Benefits and Challenges.

the same time,[15] precision agriculture includes a wide range of tools that growers need to master to get the most out of their production.

Indeed, it is important to support farmers in their transition to precision farming through the development of skills among farmers so that they can use precision farming technologies effectively; the creation of infrastructures such as learning centers and workspaces to support farmers in their transition to precision farming. Digitization and artificial intelligence (AI) play a key role in the development of precision agriculture. This approach aims to optimize the use of resources and improve agricultural productivity through advanced technologies. Digitization refers to the integration of digital technologies in various sectors. In agriculture, this includes the use of sensors, drones, geographic information systems (GIS) and farm management software. These technologies enable the collection of real-time data on crop conditions, climate and soil conditions, providing a rich database for informed decision-making (Zhang et al., 2019) .[16]

[17]Artificial intelligence, a sub-discipline of computer science, enables complex data to be processed and analyzed. In precision agriculture, AI is used to analyze large volumes of data collected by sensors and drones to identify patterns, predict yields and detect disease (Liakos et

[15] Nature Sustainability (2020). Precision agriculture for sustainable development.

[16] Zhang, C.et al. (2019). Digital agriculture: Transforming agriculture through digital technologies.

[17] Kamilaris and Prenafeta-Bolu, (2018). Deep Learning in Agriculture: A Survey. Computers and Electronics in Agriculture.

al., 2018). Machine learning algorithms can help predict crop water and nutrient requirements, enabling targeted application of irrigation and fertilizers (Kamilaris & Prenafeta-Boldú, 2018).

AI can also be integrated into autonomous agricultural machinery, such as tractors and harvesting systems, to work more efficiently and reduce labor costs. The synergy between digitization, AI and precision agriculture achieves several goals. By combining AI with digital data, farmers can optimize cultivation practices, reducing waste and increasing productivity. These technologies promote more sustainable use of resources by enabling more precise management of inputs, thus helping to protect the environment[18] . The integration of AI in agriculture enables better adaptation to climatic variability by predicting future conditions and advising on appropriate practices. Digitization and artificial intelligence are essential catalysts for the development of precision agriculture. The integration of these technologies not only makes it possible to optimize agricultural production, but also to respond to the environmental and economic challenges facing the agricultural sector. As these technologies continue to evolve, their impact on agriculture and rural development is set to grow, making the sector more sustainable and resilient.

So what are the consequences of digitalization on rural economies in Africa? How is artificial intelligence transforming precision agriculture in these economies? The present study and

[18] Nassour. A. Barnett ;(2020). Sustainable agriculture: integrating precision agriculture and smart technologies.

research of rather dazzling interest will enable us to analyze the impact of digitalization on African rural economies while examining the specific role of artificial intelligence in improving precision agriculture. Digitization is having a positive impact on rural economies in Africa, of course by improving productivity and profitability. The role of AI in the adoption, progress and success of precision agriculture is well established.

In our development, we will first outline digitalization in Africa and its impact on rural economies (**Chapter 1**); secondly, we will present precision agriculture in Africa (**Chapter 2**); thirdly, the role of artificial intelligence in precision agriculture (**Chapter 3**); and finally, the challenges and prospects of integrating AI into African agriculture (Chapter **4**).

Chapter 1:

Digitization in Africa and its impact on rural economies

Africa, the world's youngest and most dynamic continent, has 55% of its population living in rural areas, representing more than 600 million people (FAO, 2019)[19] . Africa's rural economy is often marked by low productivity, endemic poverty and vulnerability to climatic shocks. However, digitalization is revolutionizing Africa's rural economy by offering new opportunities for employment, trade and access to services. The challenges of digitization in Africa are many. Internet coverage is limited, infrastructure is often insufficient and digital skills are often non-existent or limited. However, the opportunities are also significant. Digitization can help improve agricultural productivity, reduce costs and improve access to markets. Digitization is already having a positive impact on Africa's rural economy. Young Africans are using e-commerce platforms to sell their agricultural produce and acquire goods and services. Farmers are using mobile applications to manage their farms, cut costs and improve productivity . [20]

[19] FAO (2019).

[20] Idem.

Figure2 : How Agritech and digital culture are transforming agriculture in Uganda

[21]Digitalization is revolutionizing the rural economy in Africa, offering new opportunities for employment, trade and access to services. Young Africans are using e-commerce platforms to sell their agricultural produce and acquire goods and services. Farmers are using mobile applications to manage their farms, cut costs and improve productivity. This has improved agricultural productivity in Africa by enabling farmers to benefit from online advice and training. Geographic Information Systems (GIS) also enable farmland to be mapped and crops to be planned more efficiently. Digitization has also enabled African farmers to access global markets. E-commerce platforms such as Alibaba and Amazon have created opportunities for African farmers to sell their agricultural produce on a global scale.

[21] African Development Bank (2020). Digitalisation in Africa.

[22]Digitization has created new jobs in the agricultural and service sectors. Young Africans are working in logistics, marketing and agricultural service provision. Digitization has also cut costs for African farmers. Mobile applications enable farmers to manage their farms remotely, reducing labor costs. It has a significant economic impact on Africa's rural economy. According to a study by the International Fund for Agricultural Development (IFAD), digitalization can increase agricultural production by 20% to 30% in African countries (IFAD, 2020)[23] . A powerful lever for Africa's rural economy. It offers new opportunities for employment, trade and access to services. However, it is essential to take into account the challenges associated with digitalization and to develop solid infrastructures to guarantee Internet coverage and network stability.

[24]Some success stories include :

- **M-Pesa**: a mobile money transfer platform that has revolutionized payments in Africa.
- **Agritech**: a platform enabling farmers to buy and sell agricultural products online.
- **FarmDrive**: a platform enabling farmers to purchase credit and financial services.

Digitization must be accompanied by training and awareness-raising among rural populations, so that they can make effective use

[22] Id.

[23] IFAD(2020). Digital Development in Africa: A Review of the evidence.

[24] FAO (2019). The state of food security and Nutrition in the world.

of these technologies. It is also essential to develop solid infrastructures to guarantee Internet coverage and network stability.

1-1-Definition and key concepts

There are a number of key concepts to be defined in the context of this study, and the focus will be on the notion of digitalization, while outlining the various components inherent in and linked to it, and those of the rural economy.

1-1-1-Digitization: definition and components

[25]Digitization is a term used to describe the process of integrating digital technologies into all aspects of an organization or company. This goes beyond simple digitization, which involves the conversion of analog data into digital format, to include the transformation of processes, services and business models through advanced technological solutions. digitization can be defined as the process by which companies, governments and even individuals integrate digital technologies into their daily operations and interactions. [26]According to the McKinsey Institute report (2016), digitalization involves the use of digital solutions to improve or transform business models, operational strategies and the way organizations interact with their customers. The term "digitization" comes from the English word

[25] General Data Protection Regulation (GDPR).

[26] Minkinsey Institute Report, (2016).

"digitalization", which emerged in the context of operational efficiency and digital transformation. Historically, digitalization evolved with the advent of computing and the Internet in the 1990s, but it came into its own with the explosion of mobile technologies, artificial intelligence and big data in the 21st century. There's a difference between digitalization and digitization.

[27]Scanning refers strictly to the conversion of physical data and information into digital formats. For example, scanning a document to convert it into a PDF file constitutes digitization. Digitization, on the other hand, encompasses broader processes. Not only does it transform the way information is stored and managed, it also fundamentally changes the structure and operations of companies. For example, the transition from paper-based customer management to an online customer relationship management (CRM) system represents digitization. [28]Digitization can be legally defined in several ways, depending on the context in which it is used. In general, digitization refers to the process by which analog information, processes or services are transformed into digital formats. Here's a more precise definition in a legal context: **"Digitization is the process of converting analog information, documents and activities into digital formats, enabling them to be managed, stored and transmitted via electronic devices. This process involves the use of digital technologies to automate business processes, improve**

[27] Directive 2014 /24/EU.
[28] Idem.

efficiency, transparency and traceability, while respecting legal and regulatory frameworks that protect personal data and electronic transactions."[29]

[30]For sociologists, the definition of digitalization can focus on the social, cultural and behavioral aspects of the integration of digital technologies into everyday life and into organizations. Here's a definition developed from sociological perspectives:

"Digitization is a socio-cultural process that involves the integration and adoption of digital technologies in all aspects of human life, leading to transformations in social interactions, cultural practices and organizational structures. This phenomenon engenders changes in power dynamics, modes of communication, and social relationships, while raising ethical issues and inequalities of access to digital resources."[31]

Digitalization is primarily rooted in technological innovation. The ability of companies to adopt and integrate new technologies, such as cloud computing, artificial intelligence, the Internet of Things (IoT) and blockchain, defines the depth of their digital transformation. [32]A

---

[29] Loi informatique et Libertés of 1978. In France, this law has been amended to incorporate the RGPD and provides a legal framework for digital data management, reinforcing the notion of digitalization in a legal context.

[30] Manuel Castells: In his 1916 book "The Network Society", Castells proposes the idea that digitalization modifies social structures by fostering network configurations that redefine human interactions and systems of power. In his view, the information age is fundamentally transforming social relations through digital technologies.

[31] Nicholas Luhmann; in his theory of social systems; and Sherry Turkle in "Alone Togethers".

[32] A Capgemini Study (2018).

study by Capgemini (2018) highlights that companies that invest in advanced technologies achieve significant productivity gains.

Firstly, **at organizational level,** digitalization requires a reorganization of internal structures and business processes. This may involve adopting agile methods, revising workflows or setting up cross-functional teams. Thus, the success of digitalization depends on the ability of organizations to change their corporate culture and be flexible in their operations.

On the social front, digitalization can influence human interaction and behavior. The development of social networks and digital platforms has transformed the way people communicate, consume information and connect with each other. This also has implications for privacy and data security. [33]Digitization has transformed the way we work, with significant impacts on professions, skills and working patterns. Digital technologies have created new jobs, particularly in innovation, learning and systems maintenance. However, digitalization has also threatened some traditional professions, such as office workers and manufacturing workers.

Digitalization has created new social inequalities, particularly in terms of access to the Internet and digital technologies. People with limited access to the Internet or digital technologies are often able to develop fewer skills and opportunities than those with free, unlimited access[34] . It has transformed interpersonal relationships, with impacts

---

[33] European Network and Cybersecurity Authority(ENISA)(2019). Cybersecurity and the fucture of work.

[34] International Labor Organization (ILO) (2020).

on communication, cooperation and collaboration. Digital technologies have created new opportunities for social interaction, but have also threatened certain forms of human communication[35] . Digitization also has important implications for mental health. People exposed to digital technologies for too long can develop mental health disorders, such as anxiety, stress and depression. Finally, digitalization has transformed the way we exercise our citizenship, notably in terms of political participation, access to information and civic rights. Digital technologies have created new opportunities for citizens to engage in political life and defend their rights.[36] Digitization has significant social impacts, particularly in terms of transformations in work, inequalities, interpersonal relations, mental health and citizenship. Taking these implications into account is essential if we are to promote a fairer, more sustainable society.

Secondly, **on an economic scale**, digitalization has a major impact on productivity and efficiency. It offers companies the opportunity to streamline operations, reduce costs and improve the quality of services offered. [37]According to a study by PwC (2017), companies that adopt digital technologies can increase their sales by 15-30%. Digitization, or the integration of digital technologies into all aspects of society, has transformed the global economic landscape. Whether in large corporations or small SMEs, this technological evolution has brought about profound changes in the way companies

[35] Pew Research Center (2020).

[36] United Nations Development Programme(UNDP) (2020). Human development report.

[37] A study by PwC(2017),

operate, interact with their customers, and compete. This spotlight looks at both the benefits and challenges that digitalization brings to the business world.

One of the main benefits of digitalization is the increase in business productivity. The integration of digital systems enables repetitive and time-consuming tasks to be automated, resulting in significant time savings. [38]According to a study by McKinsey (2019), companies that adopt digital technologies can increase their productivity by 20-30%.

In addition, digital tools such as big data and artificial intelligence enable companies to analyze massive volumes of data to improve their decision-making processes. For example, major retail companies such as Amazon use algorithms to personalize their offer and optimize their inventories, thereby increasing their operational efficiency (Meyer, 2020)[39] . Digitization has also catalyzed innovation at all levels. The digital startups, also known as "unicorns", often emerge with innovative ideas that have the potential to transform existing markets or create new ones. For example, the rise of sharing platforms like Uber and Airbnb has redefined the transportation and accommodation sectors. Established companies must also innovate to remain competitive. Digital transformation is pushing them to develop new products and services tailored to the changing needs of

[38] A Mckinsey Study(2019).

[39] A Meyer study (2020).

consumers. [40]The PwC report (2021) highlights that 61% of CEOs plan to accelerate their digital transformation to drive innovation.

While digitalization offers new opportunities, it also presents challenges in terms of employment. Tasks previously carried out by humans are now automated, which can lead to job losses in certain sectors. A study by the International Labour Organization (ILO) (2020)[41] indicates that almost 40% of jobs in developed countries are threatened by automation.

However, this must also be weighed against the creation of new professions linked to digital technologies. Positions such as data scientist, cyber security expert or digital marketing specialists are becoming increasingly in demand. The McKinsey report (2021) points out that a transition to jobs requiring technical skills could compensate, to a certain extent, for job losses due to automation. Digitization can also have a significant impact on sustainable development and corporate social responsibility. Through data collection and analysis, companies can track and reduce their carbon footprint, optimize their supply chain, and promote sustainable practices (GeSI, 2021)[42] . For example, the digitization of energy resource management systems contributes to more responsible consumption and reduced waste.

However, the digital transition also raises concerns about equity and access to technology. The digital divide between developed and

---

[40] Id.

[41] International Labour Organization (ILO) (2020).

[42] Id.

developing countries could widen if measures are not taken to guarantee access to technology for all. [43]According to a UNESCO report (2020), every individual must have access to digital tools if they are not to be left behind in this era of digitalization. Digitization has a significant influence on the economy, bringing productivity gains, fostering innovation, and changing the employment landscape. It presents impressive opportunities, but also challenges that need to be addressed. To reap the full economic benefits of digitalization, it is essential that governments, businesses and civil society work together to ensure an inclusive and sustainable transition. Ultimately, our economic future will depend on our ability to embrace technological change while remaining mindful of social and environmental impacts.

Finally, **from an environmental point of view**, digitalization can also contribute to sustainable and environmentally-friendly initiatives. Thanks to digital solutions, organizations can reduce their carbon footprint by optimizing resources. For example, IoT-based energy management systems enable more efficient energy consumption. [44]Digitization, which refers to the increasing integration of digital technologies into human activities, has a significant impact on many areas, including the environment. While some digital technologies can contribute to more sustainable practices, others can also exacerbate environmental problems. Digitization enables more efficient management of natural resources. Thanks to technologies

[43] UNESCO report (2020).
[44] Id.

such as the Internet of Things (IoT), companies can collect and analyze real-time data on their consumption of energy, water and raw materials. This helps identify inefficiencies and reduce waste. For example, automated energy management systems can reduce energy consumption in commercial buildings by up to 30% (United Nations, 2019) . [45] digitalization also facilitates the implementation of circular business models, where products are designed to be reused, repaired or recycled. Digital platforms enable information on waste and materials to be shared, encouraging companies to recycle more. A report by the Ellen MacArthur Foundation (2021) points out that the use of digital technologies can double the recycling rate in circular economies. Digital tools can help reduce carbon emissions.

For example, telecommuting, facilitated by digital platforms, has reduced commuting, which in turn reduces air pollution and individual carbon footprints. [46]A study conducted by Harvard Business School (2020) showed that switching to telecommuting during the COVID-19 pandemic led to a significant reduction in carbon emissions in several major cities.

Despite the potential benefits, digitization also has negative environmental consequences. Data centers, which store and process huge quantities of data, consume a considerable amount of energy. According to a report by the International Energy Agency (IEA) (2020)[47] , data centers accounted for around 1% of global electricity

[45] United Nations (2019).

[46] A study by Havard Business School (2020).

[47] IEA. (2020).

consumption in 2018, a figure that is set to rise with the continued growth of digital data. Digitization is also driving rapid obsolescence of electronic devices. Smartphones, computers and other digital devices are becoming obsolete due to new technologies and frequent updates, generating a huge amount of e-waste.[48] According to the Global E-waste Monitor (2020) report, nearly 53.6 million tons of e-waste were generated in 2019, a figure that is expected to rise to 74 million tons by 2030. The production of digital devices relies on limited natural resources, such as the rare metals needed to manufacture electronic components. The extraction of these minerals has a significant environmental impact, including deforestation, soil and water pollution, and loss of biodiversity (World Economic Forum, 2021). [49]

To minimize the negative environmental impacts of digitalization, it is essential to develop more sustainable solutions. Companies and governments can invest in renewable energies to power data centers, and promote technologies that design more durable and repairable devices. Sustainability awareness and training for digital professionals are also crucial. With the right education, digital players can make informed decisions to reduce the environmental impact of their activities. Certification programs to support sustainable practices can be established in the technology sector.

---

[48] Global E-Waste Monitor report (2020).

[49] World economic Forums 2021.

Finally, to meet the environmental challenges posed by digitization, partnerships between companies, governments and non-governmental organizations (NGOs) are needed. These collaborations can foster innovation and the sharing of best practices for a more sustainable digitization. Digitization represents both an opportunity and a challenge for the environment. While it can lead to more sustainable practices, it can also exacerbate certain environmental problems, particularly with regard to energy consumption and waste management. To maximize the benefits of digitalization while minimizing its negative impacts, it is crucial to adopt a balanced approach that incorporates sustainable solutions, increased awareness and cross-sector collaboration. The future of digitization must be designed with environmental sustainability in mind, so that we can benefit from technological advances while preserving our planet for future generations[50] . Digitization offers companies the opportunity to better understand their customers through data analysis and artificial intelligence. This enables them to personalize offers, improve customer service and optimize marketing processes. Digitization, the integration of digital technologies into business processes, has radically transformed the business landscape. From the way companies interact with their customers to the way they manage their internal operations, the impacts of digitalization are profound and multifaceted.

[50] Id.

One of the main benefits of digitalization is the ability to offer a personalized customer experience. Advances in big data and data analytics enable companies to collect and analyze customer behaviors and preferences. For example, companies like Netflix and Amazon use recommendation algorithms to suggest content and products based on purchase and viewing histories. [51]A McKinsey report (2021) mentions that 71% of consumers expect a personalized experience, prompting companies to adopt advanced segmentation techniques. Implementing chatbots and virtual assistants also facilitates customer engagement, offering instant responses and improving overall satisfaction. Digitization also enables companies to make their services and products accessible 24 hours a day. Thanks to e-commerce and mobile applications, customers can buy products anytime, anywhere. [52]According to a study by Statista (2022), online sales reached $4.9 trillion in 2021 and are set to continue growing. This shows the importance of digitization in meeting the expectations of modern customers. Digitization has also led to the automation of many operational processes. Companies are adopting technologies such as robotics, artificial intelligence and machine learning to automate repetitive tasks. [53]A report by PwC (2020) indicates that 45% of tasks could be automated, which could lead to substantial cost and time savings for companies.

---

[51] Id.

[52] A study by Statista 2022.

[53] Id.

Automation also improves the accuracy of operations. For example, the implementation of automated inventory management systems reduces human error, optimizes inventory and improves the supply chain. By analyzing data in real time, digitalization helps companies to make more informed decisions. Analytics tools enable managers to leverage data to establish key performance indicators (KPIs), identify trends and predict future results. A report by Deloitte (2021) points out that companies that design their strategies around analytical data are 5 times more likely to make more effective decisions. Digitization also stimulates innovation, enabling companies to develop new products and services. Digital technologies facilitate research and development, rapid prototyping and market testing. For example, technology companies are adopting agile methodologies that encourage rapid iteration and adaptation to changing customer needs. A report by Capgemini (2019) states that companies with a strong digital culture are 3 times more likely to succeed in innovation.

With digitalization, companies can expand beyond geographical boundaries. Digital platforms make it possible to reach customers worldwide, without the traditional constraints of physical sales. As a result, many SMEs have successfully expanded internationally through digital channels. [54]The publication by Fernanda et al (2020) points out that 78% of SMEs that have adopted digitalization have experienced a significant increase in sales. One of the main challenges of digitalization is data security. As companies collect and store

[54] Fernanda et al.2020. The impact of digitalization on SMes.

increasing volumes of sensitive data, they become prime targets for cyber attacks. [55]A report by Cyber security Ventures (2021) predicts that by 2025, global costs associated with cyber crimes will reach $10.5 trillion. To manage this risk, companies need to invest in robust cyber security solutions and train their employees in security best practices. Digitization often involves cultural and organizational changes that may meet with resistance. Employees may fear losing their jobs to automation, or be reluctant to adopt new technologies. Change management is crucial to successful digital transformation.[56] A report by McKinsey (2020) indicates that companies that implement effective change management strategies see a 70% increase in the likelihood of successful transformation. The implications of digitalization for companies are vast and multifaceted, touching every aspect of their operations, from customer experience and marketing strategy to innovation and operational efficiency.

While digitalization offers countless opportunities for improvement and growth, it also presents considerable challenges. To successfully navigate this ever-evolving landscape, companies must be prepared to embrace change, invest in advanced technologies and focus on cybersecurity. Ultimately, those who successfully integrate digitalization strategically will be the ones to thrive in the digital economy.

[55] Report by cybersecurity Ventures (2021).

[56] Id.

Speaking of the implications of digitalization for education, digitalization has enabled the emergence of e-learning and MOOCs (massive open online courses). It has also changed the way teachers interact with students, using digital tools to facilitate learning. On a societal level, digitalization is having an impact on interpersonal relations, work and daily life . It has encouraged the emergence of new forms of work, such as freelancing and telecommuting. However, it also poses challenges in terms of employability and digital inequality. One of the main challenges associated with digitalization is digital inequality, which refers to the disparity in access to and skills in digital technologies. [57]According to the International Telecommunication Union (ITU, 2020), there are still millions of people in the world without access to the Internet, exacerbating social and economic inequalities. Digitization also exposes organizations to increased cybersecurity risks. Data breaches have cost companies billions and endangered users' privacy. Managing data security risks is therefore becoming a key priority.[58] And resistance to change within organizations is a major obstacle to the successful implementation of digitalization. Employees may fear for their jobs or be reluctant to learn new skills. This requires training efforts and change management initiatives. Digitization is a complex, multi-dimensional process that is radically transforming companies, institutions and society as a whole. While it offers major opportunities for progress and

[57] International Telecommunication Union (IUT, 2020).

[58] Idem.

innovation, it also poses considerable challenges that need to be tackled proactively.

In an increasingly digital world, it's essential for individuals, businesses and governments to embark on a thoughtful, inclusive transformation that maximizes the benefits while minimizing the drawbacks.

In an increasingly connected world, it's essential to understand the various components of this phenomenon. We have **digital technologies; data**; **business processes; organizational culture; user experience; security and compliance; leadership and strategy.**

Digital technologies are the foundation of digitalization. These include IT infrastructures, comprising the hardware (servers, computers, networks) and software that facilitate the storage, processing and transmission of data. [59]According to *Manuel Castells* (1996), the information society relies on a robust technological infrastructure that enables unprecedented connectivity and accessibility. Cloud computing has enabled companies to store and access their data and applications via the Internet. This technology offers increased flexibility, enabling companies to rapidly scale their operations. [60]According to *Armbrust et al (2010)*, cloud computing is transforming the IT landscape by offering solutions that reduce

[59] Id.

[60] Armbrust et al (2010).

operating costs while increasing efficiency. Then there's the Internet of Things (IoT). The Internet of Things connects physical devices to the Internet, enabling data to be collected and exchanged. This interconnection facilitates process automation and real-time decision-making. [61]*Gartner* (2017) estimates that billions of devices will be connected within the next few years, which will have a significant impact on how businesses operate.

As for data, it plays a central role in digitalization. It includes Big Data, which refers to the explosion in data volumes generated by digital interactions. This data has the potential to provide valuable insights for businesses. [62]*McKinsey (2011)* points out that companies that harness Big Data can make more informed and strategic decisions. Then there's data analytics, and to learn from massive data, companies are using advanced analytics techniques, such as artificial intelligence and machine learning. These technologies can transform raw data into actionable information, helping companies to anticipate trends and respond to customer needs. Business processes: digitalization is transforming them. By automating business processes using digital technologies, we can reduce manual tasks and increase efficiency. [63]According to *Brynjolfsson and McAfee (2014)*, automation can lead to a significant reduction in costs and an increase in productivity. Process reengineering involves rethinking existing workflows to

---

[61] Gartner (2017).

[62] Id.

[63] Brynjolfsson and MCAfee(2014).

maximize the effectiveness of digital technologies. [64]*Hammer and Champy (1993)* point out that reengineering can lead to significant improvements in cost, quality and lead time.

Organizational culture plays a crucial role in the success of digitalization. A cultural transformation is underway. To take full advantage of digital technologies, companies need to adopt a culture of innovation and collaboration. According to *Kotter (1996)*, successful transformation requires organization-wide commitment, fostering a mindset open to change. Digital skills are essential for navigating an increasingly technologized environment. *Hase and Wilson (2003)* argue that continuous training and skills development must be priorities for organizations wishing to successfully digitalize.

User experience (XU) is another fundamental component of digitalization, made up of user-centered design.

User-centered design aims to create intuitive digital interfaces and experiences that facilitate interaction with systems. [65]*Norman (2013)* emphasizes the importance of thoughtful design to improve customer satisfaction and operational efficiency. Companies also need to develop customer engagement strategies via digital channels. This includes using social networks, mobile applications and e-mail to create lasting relationships with customers. Issues of security and compliance are unavoidable in digitalization. With cyberthreats on the rise, cybersecurity is essential to protect organizations' sensitive data

[64] Hammer and Champy 1993.

[65] Id.

and systems. [66]According to *Gartner (2020)*, companies need to invest in cybersecurity solutions to guarantee the confidentiality and integrity of information. Alignment with laws and regulations, such as RGPD in Europe, is essential for data protection. Companies must ensure they comply with these regulations to avoid sanctions and preserve customer trust.

Finally, strategic leadership is a key component of digitalization. Developing a clear digital strategy is crucial to guiding the digitalization of companies. [67]*Smit et al (2016)* note that leadership support is fundamental to catalyze change within the organization. To achieve a successful digital transformation, it is important to support employees in the change process. Effective change management approaches minimize resistance and facilitate the adoption of new technologies.

Digitization is a complex phenomenon encompassing various interconnected components. From technology infrastructure to business processes, organizational culture and user experience, each element plays a crucial role in the success of a digital transformation. To effectively navigate this changing landscape, companies need to adopt a holistic approach that integrates all these aspects while keeping an eye on security and compliance. Leadership and strategic

---

[66] Id.

[67]Smit ,P.J. et al.(2016). The digital Enterprise ; Getting started with digital transformation

vision are also essential in guiding organizations towards a prosperous digital future.

1-1-2- Rural economy in Africa: characteristics and challenges

Africa's rural economy plays an essential role in the continent's overall development. Despite persistent challenges such as poverty, political instability and climate change, rural areas represent a significant share of the population and economy in many African countries. The rural economy encompasses a diversity of livelihoods and is not limited to the agricultural sector and the production of primary products[68] . Broadly speaking, it is possible to divide the rural economy into two categories, namely the agricultural or farming sector, which produces primary products, and the non-agricultural or non-farming sector, which includes all non-agricultural economic activities that generate income for rural households[69] . Non-agricultural activities are therefore highly heterogeneous, and include mining, manufacturing, services, construction, trade, tourism, transport and financial, personal and public services. Economic diversification of the rural economy refers to the shift from agricultural to non-agricultural activities in rural areas, or to the

[68] J.R.Davis and D.Bezemer :Key emerging and conceptural issues in the development of the RNFE in developing countries and transition economies, NRI report no 2755, Chatham Maritime(kent) , Natural Resources institutes of the University of Greenwith, DFID and World Bank, July 2003.

[69] J.R.Davies :The rural Non-Farm Economy, livelihoods and theirdiversification : Issues and options, NRI report no 2753,Chatham Maritime(kent), Naturelle Ressources institutes of the University of Greenwith,DFID and World Bank, July 2003.

diversification of economic performance in these sectors. There is strong evidence that agricultural growth and productivity are important drivers for such a transformation[70] . Increasing productivity in the agricultural sector has considerable multiplier effects, stimulating other sectors of the rural economy. In areas where there has been solid growth in the agricultural sector, the rural non-farm economy has also developed rapidly, leading to higher incomes and a degree of stability for rural populations[71] . Studies on growth linkages suggest that every dollar of value added in agriculture generates between 0.6 and 0.8 USD of additional income in the rural non-farm economy in Asia, and between 0.3 and 0.5 USD in Africa and Latin America[72] . Today, the effects of economic diversification of rural economies through the growth of the rural non-farm economy as a response to a productive agricultural sector are increasingly recognized as an important driver of structural transformation in developing countries .[73]

As part of the economic diversification of rural areas, it is important to create production and consumption links between sectors,

---

[70] S.Haggblade,P.Hazell and T.Reardon: The rural Non-farm Economy: Prospects for Growth and Poverty Reduction, in World Development, October 2010, volume 38, number 10 , pages 1429-1441.

[71] Idem.

[72] S.Haggblade.P.Hazell and P.Dorosh: "Sectoral growth linkages between agriculture and the rural non-farm economy", in Haggblade, Hazell and Reardon(eds.): Transforming the rural non-farm Economy, Baltimore, johns Hopkins University Press, 2007,pages 141-182.

[73] ILO: "The importance of agriculture", in World Employment Report 2004-05 (Geneva, 2005), pages 127-182.

and to ensure their sustainability[74] . Agricultural and non-agricultural economies are linked through production activities, but also indirectly through income or investment. Production links can be forged when growth in the agricultural sector leads to an expansion of inputs and services in the non-agricultural sector, or when activities in the non-agricultural sector - such as agricultural processing and distribution which depend on agricultural inputs - increase demand for agricultural products[75] . "The importance of agriculture", in *World Employment Report 2004-05* (Geneva, 2005), pages 127-182. The growth literature identifies two main types of link between agricultural sectors. The process of economic diversification therefore depends heavily on the productivity performance of the agricultural sector, which creates demand and supply for the non-agricultural rural economy. This process is already underway in several transition and developing countries, where between 40% and 70% of rural income and employment comes from the non-agricultural sector[76] . The structure of rural employment varies between developing regions. Non-

---

[74] The growth literature identifies two main types of link between the agricultural and non-agricultural sectors: population and expenditure. Production relationships occur downstream when part of the non-agricultural sector uses agricultural outputs as inputs. Upstream production relationships occur when the non-agricultural sector supplies inputs for agricultural production, such as agrochemicals. Expenditure linkages can be divided into consumption and investment linkages: consumption linkages refer to expenditures related to household consumption, while investment linkages refer to expenditures used to finance agricultural or non-agricultural activities. B.Davis, T.Reardon, K.G.Stamoulis and P.Winters :Promoting farm/non farm linkages in developing countries ; in B.Davies .T.Reardon.k.G.Stamoulis et P,winters (dir.de publication) : Promoting farm/non farm linkages in rural development :case studies from africa and Latin America(Rome,FAO,2002), pages 1-9.

[75] Davies er Bezemer,2003, op.cit.

[76] Davis,2003, op.cit.

agricultural work employs some 48% of adult men in Latin America, the Caribbean, South Asia and the Middle East and North Africa, and some 38% in East Asia and the Pacific, but only 20% in rural sub-Saharan Africa. For women, the proportions are generally lower[77] . Patterns of rural employment in developing countries show that most households do not specialize in agricultural or non-agricultural activities, but seek to derive income from a variety of agricultural and non-agricultural sources. Household participation in the non-agricultural economy may therefore be part-time or seasonal[78] . The decision by rural households to diversify their income appears to be based on "push factors", such as risk reduction, a reaction to a crisis, a desire to obtain liquidity to stabilize income flows, or on "pull factors", such as the realization of strategic opportunities or the complementarity of different productive activities .[79]

Three main strategies are put forward to realize the potential of economic diversification for poverty reduction and decent work[80] . Strengthening small farms,—which employ the majority of the

[77] ILO: "Skills for rural employment and community development", in Policy Brief on Skills Development for Employment(2014); World Bank;World Development Report 2008:Agriculture for Development (Washington DC ;2007); International Fund for Agricultural Development: The Rural Poverty Report 2011-New realities, new challenges, new opportunities for the next generation(Rome 2010).

[78] FAO, IFAD and WFP: The State of Food Insecurity in the World 2012, economic growth is necessary but not sufficient to accelerate the reduction of hunger and malnutrition (Rome, FAO,2012).

[79]C.B.Barrett, T.Reardon and P.Webb : " Nonfarm income diversification and household livelihood strategies in rural Africa : concepts, dynamics and policy implications ,in Food Policy , volume 26(2001),pages 315-331.

[80] Bruno Losch, "L'emploi en Afrique subsaharienne; miser sur l'agriculture pour prevenir les crises",in Perspective no 19(Paris, CIRAD, October 2012).

agricultural workforce, offers the greatest potential for production and job creation; they also generate the largest share of rural income. The dynamic expansion of small farms can be a factor in the development of an energetic and diversified local economy, while at the same time, small farms depend on the local economy for inputs, services and as a market for their products. On the other hand, large farms or plantations, often characterized by decent work deficits, do not necessarily need a diversified local economy to thrive, as they may have direct links with more distant urban markets.[81] Developing markets for food products by removing risks for their producers: farmers' risk-avoidance strategies, combined with the difficulties of gaining a foothold on international markets, explain the predominance of food production - mainly cereals and tubers - in producers' choices. Links with international markets, where they exist, are primarily through exports of traditional crops (cotton, peanuts, coffee), which mobilize traditional trade channels. High value-added exports are generally highly localized, linked to specialized operators and benefit only a small proportion of farms. National and sub-regional food markets remain the most accessible. They also benefit from strong, sustainable demand. The extension of food markets and the reduction of risks for producers are essential for innovation and rural diversification.[82] The implementation of such an approach must take place within the framework of territorial policies that strengthen the

[81] Id.
[82] Id.

links between the rural and urban worlds through the promotion and intensification of service functions for small towns and rural villages, often neglected in favor of larger metropolitan cities. The ILO's approach to economic diversification in rural areas focuses on building the capacity to provide technical advice on sectors with job-creation potential, and the policies needed to support the development of these sectors to create decent jobs.

The first element of this strategy - sectoral employment impact assessments - is based on extensive work carried out by the ILO using input-output tables, social accounting matrices and dynamic social accounting matrices, computable general equilibrium models and employment projection models. The methodologies have been applied in different situations and by different programs of the International Labour Office (ILO)[83] , and provide results that can underpin policy development. In these assessments, it is possible to focus on young people (including the link with child labor) and/or women. [84]The second element - policy measures to support the development of specific sectors - can take different forms. One of these is to analyze the constraints on growth in a sector through the prism of decent work, and to propose policy recommendations aimed at overcoming these limitations. A wide range of policy areas can be covered, such as skills upgrading, enterprise and cooperative creation, occupational health and safety, social security coverage, labor inspection, etc. The ILO has

[83] International Labour Office(ILO).

[84] Ditto

developed several methodologies for this kind of analysis, including Training for Rural Economic Empowerment (TREE) and Skills for Trade and Economic Diversification (STED). Another way of looking at the design of coherent policy measures is to adopt a geographical approach and develop rural employment strategies involving local stakeholders.

In practical terms, this means supporting policy planning and program development for rural employment at local level first, with a view to identifying lessons learned, as well as tools and approaches that can feed into a national strategy for rural employment. The ILO has a number of comparative advantages, as well as extensive expertise in technical fields that support the economic diversification of rural areas:

- The Organization's mandate to promote decent work at all levels and in all sectors of the rural economy (described in 1921 and confirmed in 2008 and 2011) .[85]

- Normative capabilities, as well as numerous instruments and standards to support policy-making in the rural economy, including non-agricultural sectors.

- A tripartite structure that encourages social dialogue and strengthens the skills of the social partners.

- Technical capabilities in many areas, such as local economic development, labor-intensive infrastructure investments,

[85] Id.

entrepreneurship, cooperatives, value chains, microfinance, skills and training for rural empowerment, employment policies (informal employment), occupational health and safety, social protection, labor inspection, sectoral policies and sectoral employment impact assessments.

- Partnerships with United Nations agencies and other organizations (the Food and Agriculture Organization of the United Nations (FAO)[86], the International Cooperative Alliance (ICA), the International Fund for Agricultural Development (IFAD)[87] , the World Tourism Organization (WTO), etc.) The experience gained by the ILO through its technical cooperation projects in rural areas provides a wealth of lessons for designing interventions that support economic diversification in rural areas.

Africa's rural economy is primarily based on agriculture, which employs nearly 60% of the working population in many countries. [88]According to the World Bank, around 70% of Africans live in rural areas, and more than half of these depend on agriculture for their livelihoods (World Bank, 2020). Agriculture in Africa can be divided into several categories, including subsistence farming, commercial farming and livestock breeding.

Subsistence farming is practiced by many smallholders who grow crops primarily for their own consumption. This is often the case in countries such as Malawi and Ethiopia, where farmers grow maize,

---

86 Id.

87 International Fund for Agricultural Development (IFAD).

88 World Bank 2020.

pulses and other food crops. This form of agriculture is often subject to unfavorable climatic conditions and inefficient farming techniques. Subsistence farming is a fundamental agricultural system that constitutes the main source of food for millions of people around the world, particularly in developing countries. It is characterized by farming practices designed primarily to satisfy the food needs of a family or community, rather than for sale on markets. [89]Subsistence farming is generally practiced by smallholders who cultivate land for their own consumption. This type of agriculture can be found in various parts of the world, including Africa, Asia and Latin America. Subsistence farmers often grow a variety of food crops, such as maize, rice, pulses, as well as fruit and vegetables. In sub-Saharan Africa, for example, crops such as millet, sorghum and cassava are common (FAO, 2020)[90] . Multi-cropping is often practised, enabling farmers to ensure food diversity and reduce the risk of losses due to climatic conditions or disease.

Subsistence farming generally uses traditional, often low-tech agricultural techniques. The use of chemical fertilizers and pesticides is limited, and families rely on precise cultivation methods that have been handed down from generation to generation. These practices are often adapted to local ecological conditions, but can sometimes lack profitability and efficiency (World Bank, 2021) .[91]

---

[89] Subsistence farming; www.futura-sciences.com .

[90] Id.

[91] World Bank 2021.

Subsistence farmers depend heavily on climatic conditions for food production. Droughts, floods and other extreme weather events can have devastating consequences on harvests. This dependence makes farmers vulnerable to environmental fluctuations and food crises. Subsistence farming plays a crucial role in food security, in particular in regions where formal food distribution systems are underdeveloped. It enables rural communities to meet their basic food needs and avoid famine (WFP, 2019)[92] . In some parts of Africa, such as Ethiopia or South Sudan, subsistence farming can be the first line of defense against food insecurity.

Although subsistence farming does not generate large, regular incomes, it is a source of employment for a large proportion of the rural population. Millions of people depend on subsistence farming for their survival, and it constitutes the main means of livelihood in many communities (FAO, 2019)[93] . In addition, secondary activities linked to agriculture, such as food processing and selling products on local markets, can provide additional income. Subsistence farming promotes social cohesion within rural communities. Traditional farming practices often involve collaboration between neighbors and family, strengthening community ties. Resource-sharing systems, such as bartering and agricultural associations, also encourage solidarity (IFAD, 2021) .[94]

---

[92] PAM (2019).

[93] Id.

[94] IFAD, (2021).

One of the main challenges facing subsistence farming is its vulnerability to the effects of climate change. Variations in rainfall, rising temperatures and extreme weather events threaten production and food security. [95]According to the FAO (2020), nearly 75% of subsistence farmers in Africa are already feeling the impacts of climate change. Subsistence farmers often have limited access to financial resources, modern techniques and markets. This lack of resources limits their ability to invest in agricultural technologies and sustainable practices. In addition, access to land is often problematic, with many families lacking formal rights to the land they cultivate. Rapid population growth puts additional pressure on subsistence farming. Urbanization, which attracts many young people to the cities in search of economic opportunities, leaves a reduced number of workers for agriculture. This trend can lead to lower productivity and increased food vulnerability (UNDP, 2018)[96] . Capacity building for subsistence farmers is essential to improve their production and resilience. Training programs in sustainable farming techniques, water management systems and crop diversification strategies can help farmers better adapt to environmental change. Improving access to markets is crucial for subsistence farmers. Initiatives to develop rural infrastructure, facilitate access to credit, and set up marketing platforms can help farmers sell their produce at fair prices and increase their incomes .[97]

---

[95] Id.

[96] UNDP. (2018). Human Development indices and Indicators :2018 Statistical Update.

[97] Idem.

Integrating modern technologies into subsistence farming can also help boost productivity and sustainability. For example, the use of biotechnologies to improve crop resistance to disease and extreme weather conditions could transform this sector.

Subsistence farming remains a fundamental pillar of food security and rural development in many parts of the world. Although it faces significant challenges, including climate change and limited access to resources, opportunities exist to improve its resilience and productivity. By increasing commitment to subsistence farmers, investing in infrastructure and training, and facilitating access to markets, it is possible to strengthen this vital sector and improve the living conditions of millions of people. Commercial agriculture, on the other hand, is more developed in certain regions, where crops such as coffee, cocoa and cotton are produced for export. Countries such as Côte d'Ivoire, Ghana and Uganda are major producers of these crops. Commercial agriculture is often associated with larger farms and better integration into international markets. [98]Commercial agriculture has evolved to become one of the pillars of the global economy. Focusing on the production of crops and animals for sale on internal and external markets, this agricultural model plays a crucial role in income generation, improved food security and rural development. This editorial explores the characteristics of commercial agriculture, its role in the economy, the challenges it faces and prospects for the future.

---

[98] Id.

Commercial farming differs from subsistence farming in its aims and practices.

Commercial agriculture is primarily focused on the production of goods for sale. Farmers engage in the cultivation of products that have a market demand, such as cereals, fruits, vegetables, coffee, cocoa, and cotton. As a result, these producers must remain attentive to fluctuations in market demand (FAO, 2019)[99] . It is often characterized by large-scale farms that benefit from economies of scale. Large-scale farms can afford to invest in modern technologies, equipment and advanced management systems, increasing productivity and reducing production costs. [100]Commercial farming often relies on technological innovations. The use of genetically modified seeds, precision cultivation techniques, modern irrigation and biotechnologies can result in higher yields and greater resilience to extreme weather conditions (IFAD, 2020)[101] . It plays a vital role in many facets of the global and local economy. One of the main advantages of commercial agriculture is its ability to generate significant income. For example, products such as coffee and cocoa are major sources of income for many developing countries. [102]According to the World Bank, agricultural exports account for around 10% of the economies of many low- and middle-income countries. Commercial agriculture is an essential source of

[99] FAO 2019.

[100] www.future-agricultures.org.

[101] Id.

[102] World Bank (2021).

employment. It creates opportunities for millions of workers in farming, food processing and distribution. In Africa, for example, commercial production of crops such as cocoa, cotton and fruit contributes to rural employment and poverty reduction (FAO, 2019)[103]. Commercial agriculture stimulates the development of infrastructure, including roads, irrigation systems and storage facilities. These infrastructures are essential for the efficient transport of agricultural products to markets, thus guaranteeing market access by smallholders (IFAD, 2020).

Despite its undeniable advantages, commercial agriculture faces a number of challenges. Climate change represents one of the greatest challenges for commercial agriculture. Extreme weather conditions, droughts and floods are damaging harvests and increasing the vulnerability of farms. According to the FAO, agricultural yields could fall by 25% by 2050 if adaptation measures are not implemented[104] . The expansion of commercial agriculture has also led to the concentration of markets, where a handful of entities control the production and distribution of many products. This can reduce crop diversity and harm small farmers, who may be forced to sell at very low prices due to strong competition (World Bank, 2021)[105] . Intensive commercial farming often leads to overexploitation of natural resources, particularly water and soil. Excessive use of fertilizers and

---

[103] Id.

[104] www.fao.org.

[105] Id.

pesticides can also have adverse effects on the environment, leading to soil degradation and pollution of water sources.

To meet the challenges of commercial agriculture, the sector needs to integrate sustainable, environmentally-friendly practices. Regenerative agriculture, with its emphasis on restoring soil and ecosystem health, is a promising approach that can improve long-term productivity while preserving natural resources (FAO, 2020). Technological advances, such as precision farming and artificial intelligence , can transform commercial agriculture. These technologies enable us to better manage resources, reduce losses and optimize yields. Investment in research and development is crucial to fostering these innovations.[106] [107]It is essential to improve the resilience of agricultural systems to climate change. Policies must promote diversified farming systems, water and soil management, and financial support for farmers to adapt to new environmental realities (World Bank, 2021).

Commercial agriculture represents a significant opportunity to boost economic development, generate income and improve food security worldwide. However, it faces major challenges, including climate change, market concentration and resource sustainability. To secure its future, it is crucial to adopt sustainable practices, encourage technological innovation and strengthen farm resilience. Collaboration between governments, non-governmental

[106] IFAD. (2020).The role of agriculture in the Global Economy :improving growth and Reducing Poverty.

[107] www.enhancedif.org.

organizations, the private sector and farmers will be essential to transform commercial agriculture into a positive force for sustainable development[108] . Livestock farming is also a key element of the rural economy, in particular in regions such as the Savannah and Sahel. Livestock breeders, such as the Peuls and Touaregs, depend on grazing and transhumance. Livestock farming contributes to food security and income generation for many families.Livestock farming is a vital component of the rural economy in many parts of the world. Whether for the production of meat, milk, leather or other animal products, livestock plays a crucial role in income generation, food security and the socio-economic development of rural communities. This paper explores the importance of livestock farming in the rural economy, the challenges it faces, and future prospects for improving its sustainability and productivity. Livestock production is a major source of income for many rural households, particularly in developing countries. [109]According to the FAO (2020), nearly 1.3 billion people worldwide depend on livestock for their livelihoods. In countries like Burkina Faso and Mali, livestock farming contributes up to 40% of agricultural gross domestic product (GDP) (FAO, 2021). Farmers can sell live animals, dairy products, meat, eggs and other by-products to generate income.

Livestock farming plays an essential role in food security. Animal products, such as milk, meat and eggs, provide key nutrients including

[108] www.cairn.info.

[109] FAO (2020). Proceeding of Global conference Animals.

proteins, vitamins and minerals. In many rural areas, these products are a critical source of nutrition for families. According to the World Food Programme (WFP), animal products play a crucial role in the diets of vulnerable populations (WFP, 2019). Livestock farming is also a significant source of employment. Millions of people work in this sector, whether as breeders, processors or traders. In addition, livestock farming can help utilize local resources and boost the rural economy by creating value chains that benefit various stakeholders, including smallholders and rural communities (FAO, 2019)[110]

Climate change has a direct impact on livestock farming, particularly through extreme weather events such as droughts and floods. These events can reduce the availability of pasture and water resources, thus affecting livestock productivity. [111]According to the FAO (2021), climate change could lead to a significant reduction in yields in livestock systems, increasing the vulnerability of livestock farmers. Livestock farmers face constant threats from animal diseases, which can have devastating consequences for livestock and food security. Epidemics of diseases such as foot-and-mouth disease or peste des petits ruminants can lead to massive economic losses. Surveillance, vaccination and farmer education are essential to combat these threats (World Organisation for Animal Health, 2020) .[112]

Access to markets also represents a challenge for livestock farmers, particularly smallholders. Poor road and transport infrastructures limit

[110] FAO (2019) .The state of Food and agriculture.

[111] FAO (2021). The state of the world Land and water resources.

[112] World Organisation for animal Health (2020).

sales opportunities and increase marketing costs. In addition, price fluctuations can make economic planning and investment in sustainable livestock practices difficult (IFAD, 2020)[113] . The adoption of new technologies can improve the efficiency and sustainability of livestock farming. Practices such as sustainable intensified breeding, improved feed use, and integrated management systems can help increase productivity while reducing environmental impact (FAO, 2019). Incorporating sustainable livestock practices is essential to ensure the viability of this activity. This can include rational resource management, rotational grazing, and improved husbandry conditions. According to IFAD (2020), an integrated approach that links livestock and agriculture can increase productivity while building resilience to environmental challenges. Strengthening livestock value chains can also offer additional opportunities. This involves developing infrastructure, improving access to markets, and creating partnerships between livestock farmers, processors and distributors. Initiatives aimed at promoting local products and improving quality can play a key role in expanding markets for livestock farmers (WFP, 2019).

In Kenya, livestock farming is of great economic importance. The country is a leader in milk production in Africa, and cattle, goat and sheep farming is an essential source of income for farmers. Dairy cooperatives have been particularly effective in improving product quality and facilitating market access (Kenya Dairy Board, 2021) .[114]

---

[113] Id.

[114] Kenya Dairy Board, 2021.

In Europe, sustainable livestock systems integrated into agriculture have become the norm. European farms are adopting practices such as agroecology, which combine livestock and crops to optimize resources. Government subsidy programs also support small-scale livestock farmers to increase their resilience in the face of climate change (European Commission, 2020)[115] . Livestock farming is an essential part of the rural economy, bringing income, food security and employment to millions of people around the world. However, it faces significant challenges, such as climate change and animal diseases. The future of livestock farming will depend on the adoption of sustainable practices, the integration of innovative technologies, and the establishment of robust value chains. To ensure the sustainability of livestock farming and its role in rural economic development, it is crucial to support livestock farmers, invest in infrastructure, and promote initiatives that strengthen the resilience of rural communities.

Although Africa's rural economy has enormous potential, it faces a number of challenges: Poverty is a major problem in rural areas. According to the World Food Programme (WFP)[116] , nearly 256 million people in Africa suffer from food insecurity, which hinders their access to markets and reduces their ability to invest in their farms (WFP, 2021). This situation is exacerbated by vulnerability to climatic shocks. Climate change is having a serious impact on African

[115] European Commission .(EU) Agricultural Outlook for markets and income2020-2030.
[116] Id.

agriculture. Droughts, floods and crop diseases have become more frequent and severe. According to an FAO study, agricultural production in Africa could fall by 25% by 2050 due to climate change if adaptation measures are not taken (FAO, 2020)[117] . Limited access to markets is another major obstacle. Many rural farmers lack adequate roads, making it difficult to transport their produce to markets. In addition, the lack of storage and processing facilities reduces their ability to obtain better prices. Despite these challenges, Africa's rural economy also offers many opportunities:[118] Technological innovations have the potential to transform agriculture in Africa. Modern farming methods, mobile market communication technologies and participatory financing solutions are emerging. For example, the use of drones to monitor crops and precision farming practices can improve yields. Sustainable development and regenerative farming practices are gaining in popularity. Initiatives such as organic farming and perma-cropping can improve soil fertility and increase farm resilience to climate change. The formation of cooperatives and farming associations enables farmers to pool their resources, improve their bargaining power and gain easier access to finance and markets. Cooperatives play a crucial role in the economic stability of rural communities. [119]The rural economy in Africa is a vital sector that deserves special attention. Although a number of challenges hamper its development, there is considerable potential for

[117] Ibid.

[118] www.manutan.com .

[119] Rural economy; journals.openedition.org .

sustainable, inclusive growth. Investment in infrastructure, education, technology and institutional support can transform the rural economy and help combat poverty and food insecurity. The future of the rural economy is promising, but it requires collaboration between governments, NGOs and the private sector to realize its full potential.

1-2-The state of digitalization in Africa

Digitalization is a fundamental lever for economic and social development in Africa. With the continent at a crossroads, its digital transition offers promising prospects as well as unprecedented challenges. Internet access in Africa has grown exponentially in recent years. [120]According to a report by the International Telecommunication Union (ITU), the number of Internet users in Africa rose by 12.5% between 2019 and 2020, reaching around 495 million users in 2020 (ITU, 2021). Technological advances, such as the development of fiber optics and 4G networks, have expanded Internet coverage, even in previously underserved areas. However, access remains uneven, with marked disparities between urban and rural areas. Africa's tech startup ecosystem is in the throes of change. Technology hubs such as Kenya's Silicon Savannah, Abidjan's La Factory and Johannesburg's Ecole 42 illustrate this effervescence. [121]According to the AfriTech 2022 report, funding for African startups reached $5 billion in 2021, marking an all-time record (Deloitte,

---

[120] ITU (2021).

[121] AfriTech 2022 report.

2022)[122] . Sectors such as fintech, healthtech and agritech are experiencing unprecedented dynamism. For example, M-Pesa, a mobile money transfer service launched in Kenya, has revolutionized the financial sector by making transactions accessible to millions of unbanked people.

The rise of technology startups is one of the most striking phenomena of our time. As innovative and agile players in an ever-changing world, these emerging companies are disrupting a wide range of sectors, from finance and healthcare to education and transport. With the support of new technologies and innovation ecosystems, startups play an essential role in creating jobs, promoting economic inclusion and transforming traditional business models. The rise of technology startups is strongly supported by innovation-friendly environments. Technology hubs such as Silicon Valley in the United States, Station F in Paris, or the Binance Smart Chain in Singapore, offer ecosystems where entrepreneurs can benefit from a variety of resources, including access to financing, professional networks of experts, and incubators. According to Atomico's "State of European Tech 2021" report, the European startup ecosystem raised 100 billion euros in 2021, setting a new record and signalling a rapidly expanding dynamic (Atomico, 2021)[123] . Increasing digitization offers countless opportunities for startups. The increase in Internet access, smartphones and cloud technologies has enabled entrepreneurs to launch products and

---

[122] Deloitte (2022).

[123]Atomico.(2021). State of European Tech 2021

services with minimal costs. For example, software-as-a-service (SaaS) at has enabled startups to deliver effective solutions to a global audience without heavy investment in infrastructure. According to a study by Gartner (2021)[124] , the global SaaS market is expected to reach $800 billion by 2025. Startups benefit from a dynamic financing ecosystem. Venture capital investments, seed funds and participatory financing platforms have facilitated access to capital for many entrepreneurs. Pitch Book's Venture Capital Report (2021) reports a record increase in venture capital investment, reaching $330 billion in 2020, with continued growth observed in subsequent years (PitchBook, 2021)[125] . One of the most dynamic areas for startups is fintech. These companies innovate by offering payment, credit, wealth management and insurance solutions. Iconic examples such as Stripe, TransferWise (Wise), and Revolut have transformed the way consumers and businesses interact with financial services. According to a KPMG report, the fintech sector has raised nearly $105 billion globally by 2021, testifying to the craze for this industry (KPMG, 2022) .[126]

Startups in the healthcare sector, or "healthtech", are also expanding rapidly. The COVID-19 pandemic has accelerated the adoption of digital solutions, such as telemedicine and health monitoring applications. Companies such as Teladoc and Zocdoc are

[124] Gartner (2021) Forecast Analysis: Public cloud services, worldwide.
[125]Private Equity and Venture Capital Trends.
[126] KPMG. (2022). The Growing Global Fintech Ecosystem.

making their mark by offering e-health services that ease the pressure on traditional healthcare systems. [127]A report by Grand View Research (2021) estimates that the global healthcare technology market will reach $508.8 billion by 2029, with an annual growth rate of 28.5% between 2021 and 2029.

The education sector is also being transformed by technology startups, known as "edtech". These companies are innovating to offer e-learning solutions, course platforms and interactive educational tools. For example, companies such as Coursera and Duolingo have developed solutions that make education more accessible. According to HolonIQ, the global edtech market is expected to reach $404 billion by 2025 (HolonIQ, 2021) .[128]

Despite the positive momentum, startups face considerable challenges, not least fierce competition. In sectors such as fintech and edtech, market saturation can make it difficult to differentiate products and win market share. This requires continuous innovation and rapid adaptation to consumer needs. Startups must navigate strict regulatory environments, particularly in areas such as finance and healthcare. Ensuring compliance with regulations such as RGPD in Europe or HIPAA standards in the US can represent a considerable challenge, requiring substantial resources. Another major challenge is access to skilled talent. The growing demand for technology experts often outstrips supply, making it difficult to recruit in fields such as data

[127] Grand View Ressearch(2021) report.

[128] HoLOniQ (2021).

science and software development. According to a study by recruitment platform Hired, 77% of companies find it difficult to recruit software engineers (Hired, 2021) . [129][130]Technology startups contribute to job creation worldwide. According to a study by the European Commission, startups create 4.5 million jobs in Europe every year. This dynamic plays a crucial role in reducing unemployment and increasing employment in rapidly evolving sectors. Startups are drivers of innovation. They push established companies to reinvent themselves and improve their offering. For example, the emergence of ride-sharing platforms like Uber and Lyft has pushed the transportation industry to adopt more innovative, customer-centric practices, transforming entrenched business models.

Finally, fintech startups promote economic inclusion by making essential services, such as financial services and education, accessible to a larger number of people. In regions where traditional banking services are limited, fintech solutions make it possible to include populations previously excluded from the financial system. The rise of technology startups represents a revolution that is redefining the contours of the global economy. Although facing substantial challenges, these companies offer innovative products and services that have the potential to transform various sectors. Continued support from governments, investors and innovation ecosystems is crucial to ensure a favorable environment for startup growth. As we move

[129] Hired, (2021).
[130] Id.

towards an increasingly digital future, the influence of startups on the global economy continues to grow, promising continued transformation and shared prosperity.

Many African governments are implementing national policies and strategies to encourage digitalization. The "Digital Africa" initiative launched by the French government, as well as local initiatives such as the "Tunisie Digitale 2020" plan, aim to develop the digital economy. The Rwandan government, for example, has committed massive investment in digital technology and education, making the country a model for innovation on the continent (World Bank, 2021) .[131]

Africa, with its rich cultural diversity, abundant natural resources and dynamic population, is at a turning point in its history. To address persistent socio-economic challenges such as poverty, unemployment, gender inequality and climate change, African governments, often in collaboration with international organizations and development partners, have launched numerous initiatives and public policies. This editorial looks at three emblematic initiatives aimed at transforming the continent and promoting sustainable development: the "African Continental Free Trade Area" (ACFTA), Ethiopia's "Vision 2030" program, and Rwanda's sustainable development policy . [132][133]The African Continental Free Trade Area, created in 2018, is a major initiative to establish a single market for goods and services across the

[131] Id.

[132] The Continental African Free Trade Area (CAFTA) (2030).

[133] Idem.

African continent. The main objective of the FTAA is to expand intra-African trade and reduce the dependence of African countries on imports from outside the continent. According to the African Union, this initiative could increase intra-regional trade by 52% by 2022, and up to 83% by 2030 (African Union, 2019) .[134]

The FTAA has the potential to generate millions of jobs and stimulate economic growth on the continent. By promoting the free movement of goods, it can also boost the competitiveness of African businesses. The African Development Bank report points out that without the FTAA, intra-African trade would not exceed 12% of total trade (AfDB, 2020). Despite its potential, the implementation of the FTAA faces considerable challenges, including inadequate transport infrastructure, customs barriers and political conflicts. The World Bank warns of the risks of exacerbated inequality between the continent's rich and poor countries if corrective measures are not put in place (World Bank, 2021).

Launched in 2013, Ethiopia's "Vision 2030" policy aims to turn the country into a dynamic market economy and promote sustainable development. The initiative focuses on ambitious goals, including poverty reduction, job creation and infrastructure expansion. The Ethiopian government aspires to transform the country into a middle-income economy by 2030 (Ministry of Planning and Development, 2018)[135] . Under the aegis of Vision 2030, Ethiopia has recorded

[134] African Union 2019.

[135] Ministry of planning and development. (2018).

sustained economic growth, with an average growth rate of over 8% in recent years. Massive investments in infrastructure, such as the Grand Ethiopian Renaissance dam project, aim to meet energy needs and facilitate industrial development (Ethiopian Investment Commission, 2021). The country has also turned to key sectors, such as agriculture and textiles, to diversify its economy.

However, the country faces significant challenges, such as growing ethnic tensions, governance issues and the impact of climate change on agriculture. [136]The World Economic Forum (2021) stresses that to fully achieve the goals of Vision 2030, Ethiopia must strengthen its institutions and foster social cohesion. Rwanda is often cited as a model of development in Africa, largely thanks to its sustainable development policy integrated into the National Development Plan (NDP). The emphasis is on inclusive economic growth, environmental protection and the promotion of gender equality. Policy targets include a 30% poverty reduction by 2035 and a transition to a green economy (Rwanda Development Board, 2020). Rwanda has made considerable progress in digital infrastructure, with one of the highest Internet penetration rates in Africa. The country has also invested in renewable energy, including solar power generation to provide energy services to rural areas (Rwanda Utilities Regulatory Authority, 2021)[137] . In addition, initiatives such as the "Girinka" (one cow per family) campaign have been put in place to improve nutrition and livelihoods.

---

[136] World Economic Forum 2021.

[137] Rwanda Utilities Regulatory Authority, 2021.

However, the implementation of these sustainable policies requires considerable funding and local capacity building. In addition, concerns about freedom of expression and political diversity persist in the country, which could hamper the long-term sustainability of the progress made (Amnesty International, 2021)[138] . Government initiatives and public policies in Africa play an important role in the continent's socio-economic transformation and sustainable development. The African Continental Free Trade Area, Ethiopia's Vision 2030 program and Rwanda's sustainable development efforts illustrate the commitment of African governments to meet current challenges while exploiting growth opportunities. However, to fully realize these ambitions, it is essential to overcome challenges related to infrastructure, governance, social tensions and environmental issues. The success of these initiatives will depend on a sincere and collaborative commitment between governments, international organizations and private sector players[139] . Notwithstanding significant progress, inequalities in Internet access remain a major challenge. According to the ITU, less than 30% of the rural population in sub-Saharan Africa has access to the Internet, compared with 80% in urban areas (ITU, 2021)[140] . These disparities severely limit the ability of rural communities to participate in the digital economy and benefit from its opportunities. Digitalization is also hampered by a lack of digital skills. A World Bank study (2020) estimates that over

[138] Amnesty International (2021).

[139] Id.

[140] Id.

50% of the working population in Africa lacks the basic digital skills required to hold a job in the modern economy. This has implications for productivity and innovation, and highlights the need to accelerate training and education in this area.

In an increasingly interconnected world, digital skills have become essential for accessing economic, social and educational opportunities. Yet Africa faces a significant digital skills gap. While the continent is home to the world's fastest-growing population, challenges related to education, access to technology and infrastructure are holding back progress towards a thriving digital economy. This editorial explores the causes of the digital skills gap in Africa, its consequences on socio-economic development, and possible initiatives and solutions to remedy the situation. One of the main causes of the lack of digital skills in Africa is the inadequacy of digital infrastructures. According to the International Telecommunication Union (ITU), only 28% of the African population had access to the Internet in 2020, and inequalities in access are particularly marked between urban and rural areas (ITU, 2021). This disparity creates a digital divide that limits people's ability to acquire digital skills.

Africa's education system also suffers from a lack of integration of digital skills into school curricula. According to a UNESCO report, less than 30% of African countries have integrated digital education into their national curricula (UNESCO, 2020)[141] . This results in a lack of teaching of essential skills such as programming, Internet use and

[141] UNESCO (2020).

information management, preventing Africa's youth from developing competitive digital know-how in the marketplace. Socio-economic inequalities also exacerbate the lack of digital skills. Many households in Africa simply don't have access to technological tools such as computers and smartphones, limiting their ability to train and acquire digital skills. According to the World Bank, nearly 40% of Africa's population lives on less than $1.90 a day, making access to technology even more problematic (World Bank, 2021)[142] . The lack of digital skills has a direct impact on the continent's economic growth. [143]According to a McKinsey study, the development of the digital economy could contribute to the creation of 180 million jobs by 2030, but only if adequate training is provided and the necessary skills developed (McKinsey, 2020). Otherwise, the continent risks missing out on a crucial opportunity to use digitization as a lever for growth.

Digital skills have become a key factor in social inclusion. The inability to navigate an increasingly digital world can lead to the exclusion of many individuals from essential services such as education, health and employment. For example, the COVID-19 pandemic exacerbated these effects, highlighting inequalities in access to the digital tools needed for online education (UNICEF, 2021)[144] . Children from disadvantaged backgrounds have been particularly hard hit by these gaps. The lack of digital skills also limits innovation in Africa. Start-ups and companies can struggle to find qualified talent

[142] World Bank (2021).
[143] Id.
[144] UNICEF (2021).

for their technological projects. According to Ernst & Young's study, over 65% of African companies consider the lack of digital skills to be an obstacle to their growth (Ernst & Young, 2020). To address the lack of digital skills, several initiatives have been launched across the continent. Non-governmental organizations and local governments are working to integrate digital education into school curricula. For example, in 2018, the Rwandan government launched the "Smart Rwanda" program, which aims to equip young people with the digital skills needed for the knowledge economy (Government of Rwanda, 2018) .[145]

Many companies have also taken steps to address this gap. The "Coding for Africa" initiative connects young people with programming training and develops technological solutions locally. Similarly, partnerships between technology companies and educational institutions are becoming increasingly common, to develop training programs tailored to market needs. To improve access to technology, some initiatives are also aimed at reducing the cost of digital devices and extending Internet coverage. For example, the "Internet for All" project, in collaboration with the African Union, proposes solutions for extending Internet access to rural and remote areas (African Union, 2020)[146] . These efforts can not only increase digital skills, but also contribute to local economic development. The lack of digital skills in Africa represents a major obstacle to the

---

[145]Government of Rwanda (2018).

[146] African Union (2020). African Internet Access Initiative.

continent's socio-economic development. The challenges of infrastructure, education and socio-economic inequalities need to be tackled in an integrated way to ensure that people benefit from the opportunities offered by digitization. Through education initiatives, public-private partnerships and increased access to technology, it is possible to build an inclusive digital future for all in Africa. The collective commitment of governments, businesses and civil society will be essential to achieving this ambitious goal .[147]

Cyber security issues represent a growing challenge. As the number of Internet users increases, cyberthreats multiply, endangering the sensitive data of companies and individuals. The report by the NGO Cyber Peace Institute (2021) indicates that the number of cyber attacks in Africa increased by 60% in 2020. Governments must therefore establish robust regulatory frameworks to protect users while encouraging innovation. Collaborative initiatives between the public and private sectors have been set up in several countries to stimulate digital transformation. For example, the "Smart Africa" initiative, which brings together 30 African countries, aims to create an integrated digital market. In addition, partnerships with global technology companies support local innovation by providing resources, infrastructure and training. Financing remains a crucial element in supporting digitalization. Funds such as the African Development Fund and international organizations such as the World Bank and the African Development Bank are investing in digital-

[147] Id.

focused projects, mainly in education and technology infrastructure. The "African Development Bank" (2022) report[148] points out that investment in the digital economy could generate up to 2.7 trillion dollars by 2025.

Africa has also begun to embrace emerging technologies such as artificial intelligence and blockchain. For example, Ghana is using blockchain to improve transparency in the public sector, while AI applications are being developed for medical diagnostics in countries such as Nigeria (McKinsey, 2021). These technologies offer solutions tailored to local challenges and bear witness to entrepreneurial ingenuity on the continent.

In a world of constant technological change, Africa is emerging as a dynamic player in the adoption and adaptation of new technologies. Innovations such as artificial intelligence (AI), blockchain and the Internet of Things (IoT) offer promising prospects for the continent's economic and social development. However, challenges remain in the effective deployment of these technologies due to limited infrastructure, a lack of skills, and an insufficient regulatory framework Artificial intelligence is transforming various sectors in Africa, from agriculture to healthcare. Start-ups like **mPharma** in Ghana are using AI to improve the drug supply chain, while platforms like **2kobo** are harnessing AI to provide financial services to farmers (Kumar, 2021)[149] . AI is also being used in precision agriculture,

[148] African Development Bank Report (2022).

[149] Kumar(2021).

enabling farmers to optimize yields and minimize costs. Blockchain technology is emerging as a powerful tool for fighting corruption and improving transparency. In Kenya, initiatives such as **BitPesa** are using blockchain to facilitate cross-border payments, reducing transaction costs and improving the accessibility of financial services for small and medium-sized enterprises (SMEs) (World Bank, 2020)[150] . In addition, several African countries are exploring the use of blockchain for elections to improve the integrity of electoral processes. The Internet of Things is mainly used to improve infrastructure, healthcare and agriculture. For example, connected sensor projects to monitor water levels in agricultural fields are underway in Nigeria, enabling efficient management of water resources (Olajide, 2020). Smart cities, such as **Masdar** in Ethiopia, are also integrating IoT technologies to optimize urban services and reduce waste. The integration of emerging technologies enables the optimization of economic processes. For example, the use of blockchain in the agricultural sector helps to ensure product traceability, increasing consumer confidence and potentially raising prices for farmers (Chamberlain, 2021). AI also helps reduce operating costs and improve decision-making through accurate data analysis. Emerging technologies play a crucial role in promoting social inclusion. Blockchain-based micro platforms are opening up financing opportunities for unbanked populations, enabling them to access capital to develop their activities. What's more, AI-based digital health

[150] Id.

applications are expanding access to healthcare in rural areas, where medical services are often limited (UNICEF, 2021)[151] . In the face of challenges such as humanitarian and environmental crises, emerging technologies offer innovative solutions. For example, the use of drones to distribute food in areas of conflict or affected by natural disasters is on the increase. Such solutions help to strengthen the resilience of vulnerable communities.

Despite these advances, Africa suffers from an inadequate technological infrastructure. Limited access to the Internet and electricity are major obstacles to the implementation of emerging technologies. According to a World Bank report, some 600 million people in sub-Saharan Africa have no access to electricity (World Bank, 2020). This situation limits the capacity of local businesses to innovate and adopt technology. Developing digital skills is crucial to taking full advantage of emerging technologies. [152]According to the African Union, nearly 60% of young people in Africa face difficulties in accessing adequate digital education (African Union, 2021). This lack of skills hampers innovation and the competitiveness of African businesses in the global marketplace. The absence of clear regulatory frameworks and favorable policies is a brake on the adoption of emerging technologies. Many African countries lack laws on the protection of data and regulations concerning crypto-currencies, creating a climate of uncertainty for investors (Graham, 2021).

[151] UNICEF (2021).

[152] African Union (AU) 2021.

Proactive regulation could help create a healthy environment for technological development. Partnerships between governments and the private sector are essential to promote innovation. Initiatives such as the "Digital Africa" program launched by the African Union aim to strengthen digital skills, encourage technological entrepreneurship and promote innovation through investment in infrastructure (Digital Africa, 2020)[153] . Technology hubs and incubators play a key role in supporting innovation. Coworking spaces such as **Nairobi Garage** in Kenya and **The Aleph Network** in Nigeria offer collaborative platforms for tech startups, facilitating access to mentors, investors and the resources needed to grow (McKinsey, 2021). Capacity building is key to developing digital skills. Education and training initiatives, such as Andela's **Learn Web Development** program, aim to train local talent in web development and digital technologies. These programs create a skilled workforce capable of contributing to the burgeoning technology ecosystem (Andela, 2021) . [154][155]The adoption of emerging technologies in Africa represents a unique opportunity for socio-economic transformation. With the right strategy focused on infrastructure development, skills training and a favorable regulatory environment, the continent can finally seize the full potential of new technologies to meet its challenges and improve the living conditions of its population. The road ahead is fraught with pitfalls, but current advances and promising initiatives show that

[153] Digital Africa (2020).

[154] Andela (2021).

[155] Id.

Africa could become a leader in emerging technologies in the years to come. [156]The state of digitalization in Africa is a mixture of opportunities and challenges. While recent advances in Internet access, the emergence of innovative startups and government initiatives offer positive prospects, it is imperative to overcome obstacles such as unequal access, lack of digital skills and cyber security issues. A collective commitment between governments, businesses and civil society will be essential to realize the full potential of digitalization and ensure that its benefits are shared equitably across the continent .[157]

1-2-1- Internet and digital technology penetration rate

[158]The penetration rate of Internet and digital technologies in Africa represents a crucial issue for the continent's socio-economic development. As the world embarks on an era of digital transformation, Africa is emerging as a market both promising and facing numerous challenges.

According to the International Telecommunication Union (ITU) report for 2022, Internet penetration in Africa was around 37.5%, with significant variations between countries (ITU, 2022). For example, countries such as Nigeria, South Africa and Kenya have higher access rates, reaching 50%, 58% and 70% respectively. By contrast, countries

[156] International Telecommunication Union (IUT) (2021).
[157] CyberPeace Institute.(2021). Cybersecurity in Africa.
[158] African Development Bank.(2022). Digital Transformation in Africa.

such as the Central African Republic and Chad have access rates below 10%. This disparity is often due to economic, political and social factors. IAS and the penetration of smartphones have also played a major role in increasing access. The rise in cell phone users, estimated at 495 million in 2021, has enabled many people to connect to the internet for the first time (GSMA, 2021) .[159]

One of the major challenges facing Internet penetration in Africa is the huge gap between urban and rural areas. While large cities benefit from a well-developed telecommunications infrastructure, many rural areas lack access to high-speed Internet. According to the World Bank, only 17% of the rural population in Africa has access to the Internet, which severely limits opportunities for development and social inclusion (World Bank, 2022). Infrastructure remains a major obstacle to wider internet penetration in Africa. The lack of fiber optic cables, telecom antennas and reliable power grids is hampering the expansion of internet access, particularly in rural areas. According to a McKinsey report, around 60% of the African population has no access to electricity, which limits the ability to use digital technologies (McKinsey, 2021)[160] . The cost of internet access is also a barometer of digital penetration. The GSMA states that the average cost of a mobile data package represents around 7% of the average African's income, making access difficult for a large proportion of the population (GSMA, 2021)[161] . This situation calls for innovative

[159] GSMA(2021).

[160] McKinsey (2021).

[161] Idem.

solutions to reduce tariffs and improve the pricing architecture for digital services.

Public policies and regulations can also have a significant impact on Internet penetration. In some countries, lack of government support and restrictions on freedom of access to the Internet can hamper the expansion of digital services. For example, countries such as Ethiopia and Uganda have experienced internet outages during periods of political unrest, undermining access for citizens (Human Rights Watch, 2021)[162] . Digital technologies have a considerable impact on various economic sectors. Digitization increases efficiency, reduces costs and opens up new markets. The fintech sector, such as M-Pesa in Kenya, has shown how digital technologies can transform financial services and improve financial inclusion (World Bank, 2020)[163] . Innovation is also fostered by digital technologies, enabling startups and local businesses to grow with access to modern digital tools. They can connect to global networks, attract investment and create jobs. Online education has become a priority, especially with the COVID-19 pandemic. Many institutions have adopted e-learning platforms, enabling students to access education at a distance. The pandemic has revealed how crucial internet access is for learning, and initiatives such as "Smart Africa" aim to integrate digital solutions into the education sector (Smart Africa, 2021)[164] . The use of digital technologies in healthcare, particularly telemedicine, has improved

[162] Human Rights Watch (2021).
[163] Id.
[164] Smart Africa (2021).

access to healthcare, especially in rural areas. Mobile health applications can facilitate remote consultation with healthcare professionals, reducing inequalities in access (WHO, 2021)[165] . Efforts are being made to increase Internet penetration in Africa. Facebook's "Internet for All" project, in collaboration with governments and NGOs, aims to extend internet access to millions of people in Africa (Facebook, 2021)[166] . Similarly, the African Union has launched initiatives to improve digital infrastructure and encourage investment in the sector.

The development of educational programs aimed at training professionals in digital technologies is essential. Initiatives, such as those of the African Institute of Informatics, seek to strengthen digital skills by offering training tailored to young Africans (IAI, 2021)[167] . This will help prepare the local workforce for tomorrow's digital professions. Investing in solid, sustainable telecommunications infrastructures is crucial. Initiatives such as the deployment of fiber optic networks and satellite Internet can play a key role in improving access. Public-private partnerships could also facilitate these necessary investments. [168]Internet penetration and digital technologies in Africa are on the rise, with significant potential for economic and social development. Despite challenges such as lack of infrastructure and high costs, innovative initiatives and growing investment are

[165] WHO (2021). Global strategies on digital health.

[166] Facebook, (2021). Internet for all , A new initiative by facebook.

[167] Institut Africain d'Informatique (IAI), (2021).

[168] Idem.

paving the way for a promising digital future for the continent. By mobilizing all its resources, Africa could become a major player in the global digital economy.

1-2-2- Digital infrastructure in rural areas

Digital infrastructures play a crucial role in economic and social development worldwide, but their importance is particularly marked in rural areas. Digital connectivity is essential for accessing services such as education, healthcare and markets. However, rural areas often face unique challenges when it comes to developing this infrastructure. Digital infrastructures in rural areas are fundamental to ensuring access to information, education and health services. They also enable farmers to access information on market prices, improve their farming practices and connect with potential customers. According to the International Telecommunication Union (ITU), some 3.7 billion people worldwide have no access to the Internet, the majority in rural areas of Africa and Asia (ITU, 2021)[169] . Digital infrastructures have become of paramount importance. Not only are they the foundation of digital services, but they also play an essential role in economic development, education, health and governance. Their impact is particularly pronounced in developing countries, where limited access to digital technologies can accentuate inequalities. Digital infrastructures comprise all the technologies, equipment and systems that enable the transmission, storage and

---

[169] Id.

management of digital information. This includes telecommunications networks (fiber optic cables, antennas, etc.), data centers, as well as cloud computing services and digital applications. [170]According to the International Telecommunication Union's (ITU) 2021 report, some 4.9 billion people worldwide, or 63% of the world's population, are connected to the Internet (ITU, 2021). However, Internet access remains uneven, particularly in rural areas and developing countries. This disparity underlines the crucial importance of developing and improving digital infrastructures.

Digital infrastructures are essential drivers of economic growth. They facilitate the creation of new businesses and the development of innovation ecosystems. For example, information and communications technology (ICT) companies play a crucial role in the digital economy, contributing significantly to GDP in countries such as India and Kenya (World Bank, 2020)[171] . Access to a solid digital infrastructure also enables businesses to exploit advanced tools, such as artificial intelligence and big data, to improve their productivity and competitiveness. For example, in Kenya, the M-Pesa mobile application has revolutionized money transfer and enabled millions of people to access financial services while stimulating the local economy (GSMA, 2020) .[172]

Digital infrastructures are also an important vector for job creation. According to a McKinsey study, every billion dollars

[170] Id.

[171] Id.

[172] GSMA ;The mobile economy (2020).

invested in digital infrastructures can create around 26,000 direct and indirect jobs (McKinsey, 2019). These jobs are often in high value-added sectors, such as ICT services, cyber security and data management. Digital infrastructures play a vital role in the education sector. They enable access to online teaching resources, facilitating distance learning. UNESCO's report on education during the COVID-19 pandemic revealed that nearly 1.6 billion pupils and students were affected by school closures, highlighting the critical need for digital tools to continue education (UNESCO, 2020) .[173]

Online learning platforms such as Coursera and Duolingo offer learning opportunities to a global audience. By integrating digital education into curricula, governments and institutions can prepare students for a future where digital skills are crucial. Another crucial aspect of digital infrastructures is their ability to disseminate information. Access to reliable and relevant information is essential for the development of societies. According to a report by World Bank, access to digital information contributes to better governance and increased citizen participation (World Bank, 2021)[174] . For example, platforms for reporting problems of corruption or access to public services can strengthen civic engagement and improve transparency. Digital infrastructures can also help empower marginalized communities. They offer tools for raising awareness of social issues and mobilizing community action. Digital initiatives, such as social

[173] UNESCO (2020).

[174] World Bank 2021.

networks and messaging apps, enable citizens' voices to be heard, strengthening participatory democracy (Shirky, 2017) .[175]

Finally, digital infrastructures are playing a growing role in sustainability efforts. For example, the use of digital data and technologies to monitor the use of natural resources can contribute to more efficient management of these resources (World Economic Forum, 2021)[176] . Intelligent systems for waste management and energy use are concrete examples of the beneficial use of digital infrastructures for environmental protection.

Despite their advantages, digital infrastructures face major challenges, not least inequalities of access. Rural areas, low-income countries and disadvantaged populations are often left behind in digital development. According to the ITU, 80% of rural inhabitants in sub-Saharan Africa have no access to the Internet (ITU, 2021). Concerns about cyber security and data protection are also growing with the development of digital infrastructures. Data breaches, cyber-attacks and privacy issues pose real challenges for governments and businesses (World Economic Forum, 2021)[177] . Investment in digital security and appropriate policies are essential to ensure a secure digital environment. Digital infrastructures represent a crucial lever for economic, social and environmental development. They enable innovation, job creation, access to education and information, and contribute to community empowerment. However, the development of

[175] Shirky,(2017).The End of Audience :The changing nature of new in the digital age.
[176] World economie Forum (2021).
[177] Idem.

these infrastructures must be approached with care and exclusivity, to ensure that no one is left behind. As we move into the digital age, it is imperative to invest in initiatives that promote equitable access and robust cyber security to build a sustainable, interconnected future. At present, digital infrastructures in rural areas vary considerably from country to country, and even from region to region. In Africa, for example, rural internet penetration is less than 17% (World Bank, 2021)[178] . Countries such as Kenya and India have made progress in improving access to digital services in rural areas, investing in initiatives such as the rollout of fiber optic networks and the spread of mobile access (McKinsey, 2020) .[179]

On the other hand, many rural areas remain on the sidelines of technological advances due to insufficient telecoms infrastructure, which includes data transport networks, telecoms towers and access to the energy needed to run these technologies. According to the World Bank Group, some 600 million people in sub-Saharan Africa still do not have access to electricity, making it difficult to build effective digital infrastructures (World Bank, 2021)[180] . One of the major challenges for the development of digital infrastructures in rural areas lies in financial limitations. Initial investments in infrastructure are often high, and incomes in these regions may not justify installation and maintenance

[178] World Bank (2021).
[179] Id.
[180] Id.

costs. In addition, local governments may lack the resources to fund ambitious digital projects (Cohen & Halperin, 2020) .[181]

Telecommunications companies are often reluctant to invest in rural areas due to the low population density, which reduces the profitability of the investment. Urban areas, with a higher number of potential subscribers, are considered more lucrative, leading to an imbalance in digital infrastructure development. Rural areas can also pose technical and logistical challenges. Geography and topography can make it difficult to deploy telecoms networks. For example, mountainous regions or remote areas may require specific technologies, such as satellite or radio relays, which are often costly to implement (Fischer, 2019) .[182]

Public-private partnerships are essential for developing digital infrastructures in rural areas. Some governments are working with telecommunications companies to co-finance the deployment of networks. In India, for example, the **Digital India** program has involved partnerships with private companies to develop technological solutions in rural areas, in order to improve access to the Internet and other digital services (Ministry of Information and Communication Technology of India, 2020)[183] . Innovative solutions, such as the use of drones, satellites and community networks, have been adopted to overcome geographical and logistical obstacles. Satellite connectivity projects, such as SpaceX's **Starlink**, aim to bring

---

[181] Cohen and Halperin (2020).

[182] Fisher, (2019). Connected and unconnected.

[183] Indian Ministry of Information and Communication Technology (2020).

internet services to remote areas. These approaches can provide direct access to digital services without the need for large-scale ground infrastructure (SpaceX, 2021)[184] . Community initiatives are also important vehicles for developing digital infrastructures. Projects such as **Community Networks** in South Africa enable communities to build and manage their own telecoms networks, offering a solution to those left behind by the major access providers (Community Networks, 2020)[185] . These networks not only provide Internet access, but also build local capacity and promote self-reliance.

To ensure the sustainable development of digital infrastructure in rural areas, governments need to develop favorable policies that encourage investment. This includes tax incentives, subsidies for infrastructure projects and favorable regulations that encourage innovation in digital connectivity (OECD, 2021)[186] . Education and awareness are key to maximizing the benefits of digital infrastructures. Efforts must be made to train rural populations to use these technologies, to access digital services and to be actors in the development of their own infrastructures. Training programs and practical workshops can help overcome the digital divide (World Economic Forum, 2021)[187] . Technological advances could also be a major lever for improving digital infrastructure in rural areas. The Internet of Things (IoT) and advanced communications technologies

---

[184] SpaceX (2021).

[185] Community Networks, (2020).

[186] Organisation for Economic Co-operation and Development (OECD), (2021).

[187] Id.

such as 5G could improve the efficiency of networks and services, even in challenging environments. Digital infrastructures in rural areas represent a major challenge, but also an opportunity to significantly improve quality of life. In an increasingly digital world, it is essential to guarantee access to the Internet and digital technologies for all. By mobilizing joint efforts between governments, the private sector and local communities, Africa and other rural regions of the world can move towards an inclusive digital future that fosters economic development, education and improved living conditions .[188]

1-3-Impact of digitalization on rural economies

Digitization, or digital transformation, refers to the integration of digital technologies into all aspects of society. In the context of rural economies, it offers new opportunities to improve productivity, market access and quality of life for rural communities. However, it also poses challenges in terms of access, training and infrastructure. No more flying blind with limited data, but instead making informed decisions to successfully manage a healthy, efficient and agile farm or plantation, optimizing the use of resources: **water, pesticides, energy.** The integration of artificial intelligence (AI) into agriculture represents a major breakthrough for the sector, helping to meet increasingly pressing challenges linked to global food security. With a projected world population of around 9.2 billion by 2050, and an increase in arable land limited to just 4%, it is becoming imperative to

---

[188] Id.

optimize agricultural production methods. AI offers innovative solutions for "farming better", increasing yields while reducing costs and environmental impact.

Sustained growth in agriculture requires more than just a "green revolution". In particular, significant investment is needed in irrigation and rural infrastructure, human resource development, AI and institutions.

1-3-1-Improving market access

Digitization enables better connections between producers and consumers, facilitating access to the market. Thanks to online trading platforms and mobile applications, farmers can sell their products directly to consumers, bypassing the middleman. This can lead to higher profit margins and increased sales. For example, in countries like Kenya, platforms like M-Farm give farmers direct access to wider markets, increasing their income (Buntong et al., 2019)[189] . Digital technologies, such as IoT (Internet of Things) sensors and geographic information systems (GIS), enable farmers to better monitor and manage their crops. These tools provide real-time data on soil conditions, moisture and other environmental factors, enabling more precise resource management. According to an FAO study, the adoption of these technologies can increase agricultural yields by 10-30% in rural areas (FAO, 2020) .[190]

[189] Buntong et al (2019).
[190] FAO (2020).

1-3-2- Reducing transaction costs

Digitization is also facilitating access to financial services through mobile banking and other digital financial products, and transaction costs are being considerably reduced. Farmers can easily receive payments, save and access credit. For example, the M-Pesa mobile payment service in Kenya has contributed significantly to the financial inclusion of farmers, enabling them to invest in other economic activities (Jack and Suri, 2014) .[191]

1-3-3-Strengthening agricultural value chains

AI is transforming agriculture in a number of ways: it can collect and analyze data on climatic conditions, soil quality and other essential factors, making management easier.

Figure3 : Plant disease verification using the IA application

Thanks to seasonal forecasting systems and predictive models, farmers aided by AI will be able to manage the use of water and inputs, enabling a significant reduction in water and chemical consumption.

[191] Jack and Suri, 2014.

For example, intelligent irrigation systems adjust water supply to the specific needs of crops.

Figure4 : Using androids to list in the database

Artificial intelligence (AI) plays a crucial role in crop disease prevention, offering tools and technologies that enable early detection and effective management of phytosanitary problems. One of the main advantages of AI in crop disease prevention is its ability to detect symptoms at an early stage. These systems identify signs of disease before they spread, enabling farmers to act quickly to limit losses. Digitization helps rural communities diversify their sources of income. Farmers can engage in ancillary activities, such as rural tourism or handicrafts, using digital technologies to promote their products. This strengthens the economic resilience of communities in the face of shocks, such as fluctuations in commodity prices or natural disasters (World Bank, 2017)[192] . AI also uses real-time environmental data to monitor crop health. By analyzing factors such as soil moisture, temperature and weather conditions, algorithms can predict the risk of

[192] World Bank (2017).

disease and pest outbreaks. This enables farmers to make informed decisions on which preventive treatments to apply.

AI systems can send alerts to farmers about potential threats to their crops. By integrating predictive models, these tools provide recommendations on the best practices to adopt to prevent disease, such as the ideal time to apply phytosanitary treatments or adjust cultivation practices. AI also contributes to more efficient resource management by enabling targeted application of pesticides and fertilizers. This not only reduces costs, but also minimizes environmental impact, by avoiding excessive use of chemicals that can harm surrounding ecosystems.

Figure5 : Precision watering

It improves crop resilience in the face of sanitary threats, while optimizing resources and reducing losses. This proactive approach is essential to ensure sustainable, productive agriculture in the future.

Despite the potential benefits, limited Internet access remains a major obstacle to digitalization in rural areas. According to a report by the International Telecommunication Union (ITU), 90 % of rural populations in some parts of Africa still do not have access to the Internet (ITU, 2021)[193] . This digital divide can exacerbate economic inequalities between urban and rural areas. Agricultural value chains (AVCs) play a crucial role in economic development, food security and poverty reduction in many regions, particularly in developing countries. They encompass all activities from the production of raw materials to their processing and distribution to the end consumer. Strengthening agricultural value chains is essential for improving farm productivity, sustainability and profitability.

Agricultural value chains are important drivers of economic growth. By improving efficiency and integration within value chains, many jobs can be created, both in agriculture and in the processing and distribution sectors. For example, a World Bank study showed that investment in VSCs could boost the local economy and improve farmers' livelihoods (World Bank, 2020)[194] . Strengthening VSCs also contributes to food security by increasing production and ensuring better access to food. Efficient value chains ensure that agricultural products are available on markets at affordable prices. This reduces food waste and improves the availability of essential nutrients for vulnerable populations. A well-structured agricultural value chain can

---

193 Id.

194 World Bank (2020).

promote sustainable farming practices, thereby helping to protect the environment. For example, the adoption of agro-ecological techniques in CVAs can improve soil health and reduce the use of harmful chemicals, benefiting both biodiversity and human health (Kassam et al., 2019)[195] . Training farmers in modernized production techniques, farm management, and product quality is essential to improve their competitiveness within value chains. Training programs can be set up by NGOs, governments or private institutions, focusing on innovation, the use of digital technologies and sustainable practices (Gollin et al., 2016)[196] . Integrating digital technologies into agricultural CVAs improves traceability, collects accurate production data and facilitates transactions. Mobile applications, for example, can help farmers access agricultural advice, sell their produce and receive payments securely. This technological improvement promotes transparency and efficiency in value chains (FAO, 2020) .[197]

Collaboration between different value chain players - farmers, processors, distributors and consumers - is crucial to strengthening the effectiveness and sustainability of CVAs. Public-private partnerships can encourage initiatives that improve the quality and quantity of agricultural production. For example, public-private alliances for research and development can foster innovation (Norton et al.,

---

[195] Kassam et al (2019).The role of agroecology in strengthing agricultural value chains. Agronomy for sustainable development, 39(3),1-12.

[196] Gollin et all (2016).Agricultural Productivity Gaps in Developing Countries: Acomparison of 10 countries and their agricultural value chains , 78, 197-207

[197] Id.

2019)[198] . Farmers often face price volatility, which can affect their income and motivation to invest. Protection mechanisms, such as crop insurance schemes or forward contracts, can help mitigate these risks, but they often require institutional support and adequate infrastructure (World Bank, 2019)[199] . Financing remains one of the main obstacles to strengthening agricultural value chains. Farmers, particularly small-scale producers, often find it difficult to access affordable credit. Financial institutions need to develop tailored financial products that take into account the specificities of the agricultural sector (Agarwal et al., 2021)[200] . Agricultural value chains must also face the impacts of climate change. The variability of climatic conditions affects not only production but also the availability of resources (water, soil). It is therefore essential to integrate climate change adaptation practices into value chains (IPCC, 2021)[201] . The future of agricultural value chains lies in their ability to innovate and adopt sustainable practices. The adoption of green technologies, such as precision farming and nature-based solutions, can not only improve yields but also reduce the environmental impact of agriculture (World Resources Institute, 2020) .[202]

Strengthening agricultural value chains will also require multi-sector partnerships, involving the private sector, governments and

[198] Norton et al, (2019). Agrcultural value chains management. Journal of agricultural and resource economics, 44(1),149-179.

[199] Id.

[200] Agarwal et al (2021).

[201] IPCC, (2021).

[202] Zorld Resources Institute, (2020).

civil society. These partnerships can foster the sharing of resources and knowledge, catalyzing innovative solutions to contemporary agricultural challenges.

[203]Finally, a heightened sensitivity to farmers' needs and a strong commitment to the implementation of policies favorable to the agricultural sector will be crucial. Governments need to stimulate policies that support the sustainable development of value chains, promote research and innovation, and ensure a fair economic landscape for all players. Strengthening agricultural value chains is essential to stimulate economic development, improve food security and promote sustainability in food systems. By adopting effective strategies, while overcoming existing challenges, players in agricultural value chains will not only be able to improve their economic results, but also contribute to a more sustainable and equitable food future.

[203203] Id.

Chapter 2:

Precision farming in Africa

Precision agriculture in Africa represents an innovative approach to improving the efficiency and sustainability of agricultural practices on the continent. This approach relies on advanced technologies to optimize the use of resources, increase yields and minimize environmental impacts.

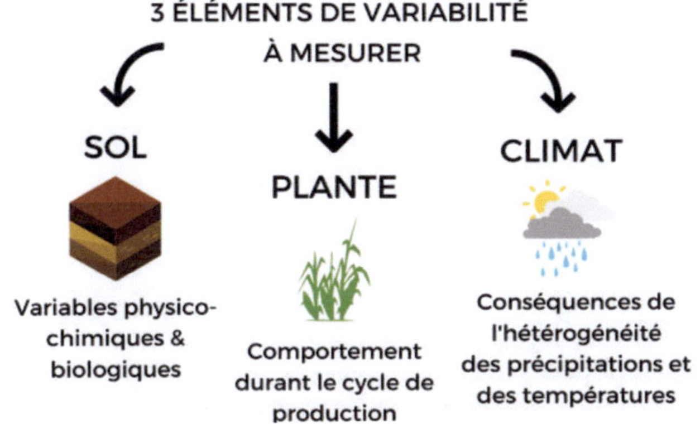

Figure6 : Precision farming concept

The concept of precision farming is based on the observation that a plot of land is not homogeneous, but heterogeneous. A plot is no longer a single entity, but a multitude of small plots. Intra-plot heterogeneity refers to the variability of soil characteristics and potential within the same plot. The new technologies are a central element of precision agriculture. The term "Agtech" or "Agritech" is now used to describe an innovative sector that combines agriculture with new technologies.

Figure7 : Agtech or Agritech

In the agricultural sector, connected objects are playing an increasingly important role. There are many different types of sensor that can be used to improve farming practices thanks to data: weather stations, drones, robots, temperature sensors for silo management or forage storage... AI predicts weather conditions and manages the associated risks, enabling farmers to adapt to climatic variations. This strengthens the resilience of crops to environmental hazards, guaranteeing stable and sustainable food production. By improving the efficiency of agricultural practices, AI helps reduce greenhouse gas emissions. For example, optimizing cultivation and harvesting processes can reduce energy consumption and transportation needs, which is essential in the fight against climate change.

2-1- Introduction to precision farming

[204]Precision agriculture, also known as precision farming (PA), is a modern approach that uses advanced technologies to observe, measure and manage variability within crops and fields. This method aims to improve the efficiency and sustainability of farming practices while increasing yields and reducing operational costs. [205]Its principles include Data Collection, characterized by the use of sensors, drones, satellites and other tools to collect data on soil conditions, irrigation requirements, crop condition, etc..; Analysis and Interpretation, characterized by data processing using specialized software to draw conclusions about crop condition and specific needs;[206] Targeted Application, characterized by agronomic decision bases on data analysis, enabling precise application of fertilizers, pesticides and water, according to specific plant needs. The technologies used in precision agriculture are :

Drones: For crop monitoring and real-time data collection.

-Soil Sensors: Measure soil moisture, temperature and nutrients.

-Farm Management Systems (FMIS): to integrate and analyze agronomic data.

---

[204] Id.

[205] Zhang, N., Wang, M., & Wang, N. (2002). Precision agriculture-A worldwide overview.

[206] Gebbers, R., & Adamchuk, V. I. (2010).

It offers a number of advantages:

- ***Optimization of resources:*** Reduced over-consumption of water, fertilizers and pesticides, contributing to more sustainable farming practices.

- ***Increased yields***: Improved productivity thanks to more precise resource management.

- ***Improved Decision Making:*** Farmers can make informed decisions based on accurate data rather than estimates .[207]

2-1-1-Definition and components of precision agriculture

Precision agriculture (PA) has become an emblematic approach to agricultural modernization in the early 21st century. It combines advanced technologies with traditional agronomic methods to achieve unprecedented levels of efficiency and sustainability. In response to the growing need for increased food production in a context of climate change and limited resources, precision agriculture offers innovative solutions. Precision agriculture is a term that encompasses a variety of practices aimed at optimizing the use of agricultural inputs and resources through an in-depth understanding of the spatial and temporal variability of fields.

[207] McBratney, A., Whelan, B., Ancev, T., & Bouma, J. (2005). Future directions of precision agriculture.

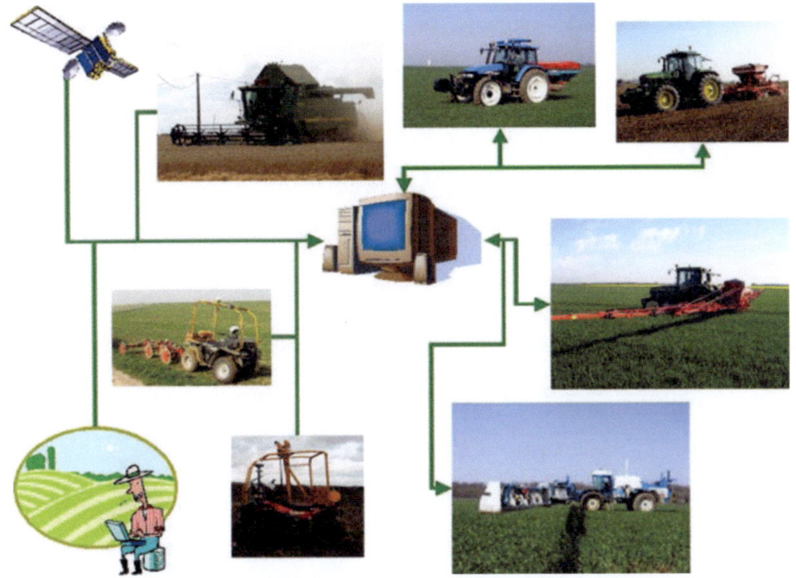

Source : https://www.agri-mag.com/2020/04/13/agriculture-de-precision-produire-plus-avec-moins/
Figure 8 : Precision agriculture organization chart

[208]According to Zhang et al. (2002), "precision agriculture refers to crop management systems that use technologies to monitor, measure and manage variability in the field to increase the profitability of agricultural production while minimizing negative effects on the environment." One of the central concepts of precision agriculture is the idea of variability. Unlike a one-size-fits-all approach that applies the same treatments to an entire field, precision agriculture recognizes that different parts of a field may require different treatments due to variations in soil composition, topography, moisture, etc. This approach also aligns with the principles of sustainable agriculture,

[208] Id.

tending to reduce chemical inputs and increase ecological sustainability (McBratney et al., 2005).[209]

2-1-2-Key technologies used [210] Precision agriculture (PA) is an innovative approach that uses advanced technologies to improve crop yields while minimizing environmental impact. By integrating data collection, analysis and management systems, PA enables farmers to make decisions based on accurate data about their fields. Key technologies include **Global Positioning Systems (GPS), UAVs and aerial imagery, IoT (Internet of Things) sensors, analysis software and Farm Management Systems (FMIS), irrigation control technologies and remote sensing systems.** [211] As a prelude, Global Positioning Systems (GPS) are at the heart of precision agriculture.

Source: https://www.agri-mag.com/2020/04/13/agriculture-de-precision-produire-plus-avec-moins/

Figure9: Producing more with less: An agricultural model reserved for the wealthy

[209] McBratney, A., Whelan, B., Ancev, T., & Bouma, J. (2005). Future directions of precision agriculture. *Soil Use and Management, 21*(4), 363-370.

[210] Id.

[211] Idem.

They enable farmers to locate their farm machinery and equipment with great precision. This technology makes it easier to map fields and plan operations such as planting, fertilizing and harvesting. GPS works by using a constellation of satellites to send signals to a receiver in the field. By measuring the time it takes for the signals to reach the receiver, it is possible to calculate its precise position. Thanks to this technology, farmers can apply inputs such as fertilizers and pesticides in a more targeted way, reducing waste and minimizing environmental impacts (Zhang et al., 2002) .[212]

The Global Positioning System (GPS) is a revolutionary technology that has transformed many sectors over the last few decades, including agriculture, transportation, aviation, marine navigation and many others. Initially designed for military applications, GPS has become widely accessible and essential in everyday life. The system is based on a constellation of satellites which transmit signals to receivers on the ground, enabling the exact position of these receivers to be determined with remarkable precision.

The development of GPS has its origins in the 1970s, when the US Department of Defense launched the NAVSTAR (Navigation Satellite Timing and Ranging) program. The initial aim was to create an accurate navigation system to enhance military capabilities, particularly in field operations. In 1978, the first GPS satellite was launched, and the system became fully operational in 1995, with a

[212] Id.

constellation of 24 satellites (Hoffman-Wellenhof et al., 2001)[213] . Over the years, GPS has evolved into a technology accessible to the general public. In 2000, the U.S. government eliminated selective jamming, giving civilian users access to increased accuracy. Today, GPS is used daily by millions of people worldwide. [214]GPS operates via a network of satellites in low-Earth orbit, which transmit radio signals containing timing and location information. A GPS receiver on the ground receives these signals from several satellites (at least four for precise positioning). Operation is based on the following steps:

Each satellite sends a signal that indicates its location and the time at which the signal was transmitted. The GPS receiver measures the time it took for the signal to reach its antenna. By multiplying this time by the speed of light, it can calculate the distance to the satellite. Using data from several satellites, the receiver can then triangulate its exact position in three dimensions (latitude, longitude and altitude) (Misra & Enge, 2006)[215] . GPS uses a global reference system called WGS84 (World Geodetic System 1984)[216] , which defines the shape of the Earth and enables uniform coordinates worldwide. This ensures accuracy and consistency in location measurements, wherever GPS is used. Various factors, such as the atmosphere and satellite positioning, can introduce errors into measurements. Techniques such as the Time

---

[213] Hoffman-Wellenhof et al, 2001.

[214] Idem.

[215] Misra, P., & Enge, P. (2006). *Global Positioning System: Signals, Measurements, and Performance*. Ganga-Jamuna Press.

[216] World Geodetic system (1984).

Position Correction System (TPCS) and the use of reference stations enable these errors to be corrected by providing real-time information (Van Dierendonck, 1996)[217] . GPS has applications in many sectors, from personal navigation to critical industrial uses. GPS is essential for road and sea navigation. GPS navigation systems help truck drivers plan their routes, avoid traffic jams and save time and fuel. In addition, GPS cargo tracking enables better supply chain management (Wang et al., 2016)[218] . Emergency services use GPS to quickly locate incidents and coordinate responses. GPS applications are also used in child locating devices and car security systems, aiding the recovery of stolen cars[219] . GPS has profoundly changed leisure activities, including camping, hiking and sea navigation. Handheld devices and hiking apps enable users to navigate with precision in unfamiliar environments, making these activities more accessible and safer (Corbett, 2010)[220] . The future of GPS looks bright, with the continued development of complementary technologies. For example, satellite augmentation systems, such as the European GALILEO system and the Global Navigation Satellite System (GNSS), promise to further improve the accuracy and reliability of location-based services. What's

---

[217] Van Dierendonck, A. J. (1996). GPS: Theory and Applications. In *Handbook of Global Navigation Satellite Systems* (pp. 15-51). Springer.

[218] Wang, J., et al. (2016). The Impact of GPS Technology on Supply Chain Management. *Journal of Modern Transportation, 24*(2), 112-120.

[219] Elmasry, I., et al. (2019). The Impact of GPS Technology on Privacy: A Review of Implications for Users and Policymakers. *Journal of Information Technology Research, 12*(3), 45-56.

[220] Corbett, J.(2010).GPS for the outdoor Enthusiast: A comprehensive guide to using GPS technology for Hiking, camping and other outdoor Activities.

more, the integration of GPS with other emerging technologies, such as the Internet of Things (IoT), could transform sectors such as smart agriculture, connected cities and disaster management[221] . The Global Positioning System has transformed the way we navigate and interact with our environment. Its development, operating principles and multiple applications illustrate its significant impact on modern life. Although it has its drawbacks, the advantages of GPS make it an essential tool in many fields. As technology continues to evolve, GPS will continue to evolve, offering even greater opportunities for the future of location data management and other innovative applications.

Secondly, drones are playing an increasingly essential role in precision agriculture because of their ability to provide detailed aerial images of fields. Equipped with high-definition cameras and multispectral sensors, drones can capture images that reveal crop health and soil variability. Farmers can use aerial imagery to monitor plant health, detect disease, assess water stress and estimate potential yields. For example, infrared imagery can help identify areas that are short of water, enabling rapid intervention (Anderson & Gaston, 2013). [222]

What's more, drones offer the advantage of reducing time and costs compared with traditional surveillance methods, making them a valuable tool for farmers. Drones and aerial imagery have revolutionized a variety of sectors, from agriculture and mapping to

[221] Hassan, H.A et al (2019). The role of GPS in Precision agriculture.

[222] Anderson and Gaston, 2013.

surveillance and disaster management. The use of these technologies enables more efficient, accurate and cost-effective data collection. Drones, also known as Unmanned Aerial Vehicles (UAVs), have their origins in early military applications. The first UAVs were developed in the 1910s for training and reconnaissance purposes. However, it was during the Vietnam War, thanks to airborne reconnaissance systems such as the Ryan Firebee, that UAVs really came into their own (Kahn et al., 2013)[223] . With the advent of digital technology and advances in sensor technology, the use of drones has expanded beyond military applications. In the 2000s, small drones equipped with cameras became accessible, enabling their use by the general public and professionals in a variety of sectors. Improved flight controllers, satellite navigation systems such as GPS, and communications technologies have also contributed to the proliferation of drones.

A typical drone consists of several essential components:

- **Structure and motor**: UAVs can be fitted with different types of motor (electric or combustion), and their construction varies according to their use (multi-copters, helicopters, etc.).
- **Sensors and payload**: The payload can include cameras, multi-spectral, thermal or LIDAR sensors, enabling a wide range of data to be collected.

[223] Kahn, D. et al. (2013). "Historical Development of Unmanned Aerial Vehicles." *The Journal of Transportation Technology*, 7, 533-548.

- **Navigation system**: Drones typically use GPS for navigation, but many recent models also incorporate non-GPS positioning systems (such as vision-based positioning) for increased accuracy in urban or cluttered environments (Kärcher et al., 2016) .[224]

Most drones can be controlled manually using a remote control, or autonomously using flight-planning software that includes waypoints. Autonomous drones use algorithms to adjust their flight to varying conditions, ensuring efficient and reliable data collection (Colomina & Molina, 2014)[225] . Drones are used to create accurate, up-to-date maps in a variety of fields, including urban planning, natural resource management and land arrangement. Drone photogrammetry is used to generate digital terrain models (DTMs) and ortho photos, useful for a variety of applications, including the planning of infrastructure projects (Chase et al., 2014)[226] . They play an essential role in environmental monitoring and disaster management. They can be deployed to assess damage after natural calamity events such as floods, forest fires or earthquakes. Thanks to aerial imagery, authorities can quickly obtain critical information for emergency response management. Drones are also used to inspect infrastructure such as bridges, power lines and pipelines.

---

[224] Kärcher, M. & Klöckner, C. (2016). "High-Resolution Mapping Using UAVs." *Journal of Geographical Systems, 18*(3), 329-345.

[225] Colomina, I., & Molina, P. (2014). "Unmanned Aerial Systems for Photogrammetry and Remote Sensing: A Review." *ISPRS Journal of Photogrammetry and Remote Sensing, 92*, 79-97.

[226] Chase, S., et al. (2014). "Evaluation of UAV-based Photogrammetry for Mapping Vegetation in a Tidal Marsh." *Remote Sensing, 6*(1), 164-182.

The use of drones for inspections reduces the need for work at height or with difficult access, thereby reducing risks for workers and cutting associated costs (Dandois & Ellis, 2010) .[227]

Source : http://grceta27.free.fr/agri_preci.php

Figure10 : Using drones for reduced inspections

Precision agriculture, enabled by AI, uses analytical data to make informed decisions about cultivation practices. This includes the targeted application of fertilizers and pesticides, which not only increases yields, but also reduces the use of harmful chemicals. By detecting diseases and pests early, farmers can intervene quickly, minimizing environmental impact. With automated irrigation systems that use sensors to determine crop water requirements, ensuring efficient use of water resources. And the use of robots and AI systems

---

[227] Dandois, J. P., & Ellis, E. C. (2010). "Remote Sensing for Biodiversity: The Use of UAVs to Evaluate Plant Species Diversity." *Ecosphere, 1*(4), 1-15.

in vertical farming enables optimized management of growing spaces, making urban agriculture more viable and productive.

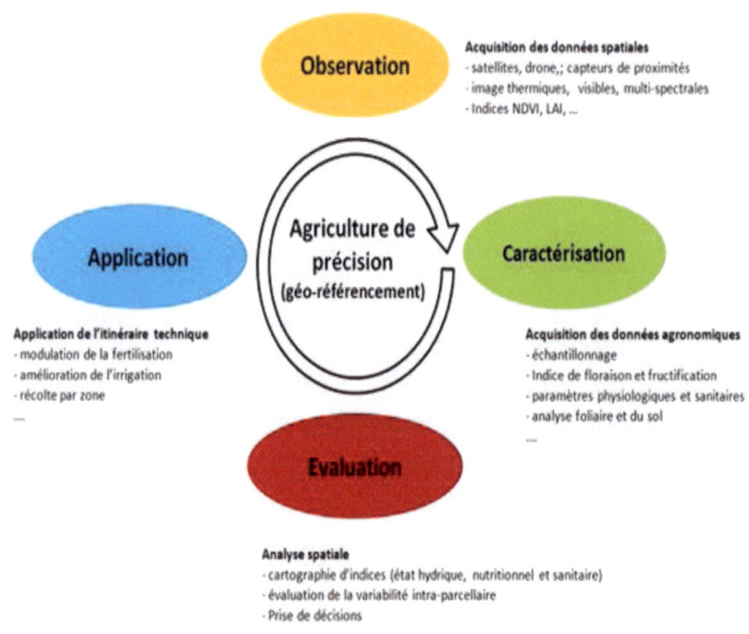

Figure11 : Referencing system

The use of drones represents a much cheaper alternative to traditional aerial imaging methods, such as the use of aircraft, allowing savings on assessment missions (Gonzalez et al., 2020). Drones can provide updated data in real time, which is particularly useful in critical situations such as emergency response. They can access hard-to-reach or dangerous areas, enabling effective monitoring and data collection in a variety of environments (Gonzalez et al., 2020)[228] . The use of drones also has its drawbacks; The use of drones is often subject to strict regulations in many countries, which can limit their operation. High-resolution aerometry raises concerns about individual privacy, such as unauthorized surveillance (Crampton, 2009). Although drones are powerful tools, they are limited by their flight autonomy and carrying capacity, which can restrict the duration of missions or the type of sensors used .[229]

The future of drones and aerial imagery looks bright, as technology continues to advance. The emergence of delivery drones, the development of more sophisticated autonomous drones and advanced communication systems are opening up new possibilities. In addition, the integration of artificial intelligence (AI) will enable collected data to be analyzed more intelligently and efficiently, often automating tasks that previously required human intervention (Murphy, 2020) . [230][231]Application trials in sectors such as logistics, transport and even space exploration augur significant changes in the way these technologies can be used in the near future. In addition, the development of clearer, global regulations around drone use could encourage wider adoption and greater social acceptability. Drones and aerial imagery represent significant advances that are transforming

---

[228] Gonzalez, A., et al. (2020). "The Economic Impact of Using UAV in Agricultural Practices." *European Journal of Remote Sensing, 53*(1), 35-50.

[229] Id.

[230] Murphy, R. (2020). "The Future of Drone Technology: Trends and Predictions." *Journal of Unmanned Vehicle Systems, 8*(2), 56-62

[231] Idem.

many sectors, offering innovative solutions, and paving the way for better resource management, emergency planning and response. While challenges remain in terms of regulation and ethical concerns, their potential to revolutionize various fields is indisputable. As technology continues to evolve, it is likely that drones will become essential tools for a wide range of applications.

[232]In addition, sensors are critical devices that collect data on various environmental and agronomic parameters. In PA, IoT (Internet of Things) sensors are widely used to measure factors such as soil moisture, temperature, pH and nutrient concentration. These sensors can be permanently installed in the soil or used mobile for spot assessments. The data collected enables farmers to make decisions based on real-time information, improving water and nutrient management (Wang et al., 2017)[233] . For example, irrigation can be optimized by automatically activating irrigation systems when soil moisture levels fall below a predefined threshold. The Internet of Things (IoT) is a technological revolution that is transforming the way we interact with our environment, by integrating intelligent devices capable of collecting, exchanging and analyzing data. IoT sensors play a central role in this transformation. They enable the collection of real-time data on various aspects of our physical world, from the domestic environment to industry, smart cities and agriculture. An IoT sensor is a device that perceives a physical phenomenon, such as temperature,

[232] www.maherelectronica.com,

[233] Wang et al 2017.

humidity, light or movement, and converts it into an electronically readable signal. These signals arc then transmitted via wireless networks to other devices or systems for analysis and action (Sundmaeker, et al., 2010)[234] . There are several types of IoT sensor, each with its own function and application:

- ***Temperature sensors***: measure ambient temperature and are used in heating, ventilation and air-conditioning (HVAC) systems.
- ***Humidity sensors***: measure humidity levels in the air, essential for agriculture and environmental management.
- ***Motion sensors:*** Detect the physical movement of an object or person and are used for security and building automation.
- ***Pressure sensors***: measure the pressure of gases or liquids and are commonly used in industrial applications.
- ***Light sensors***: Measure light intensity, used for intelligent lighting and energy management (Gubbi et al., 2013) .[235]

Most IoT sensors operate in a multi-layered architecture, from data collection to processing. This architecture includes:

- **Sensors and devices**: Collect data from the physical world and convert it into digital signals.
- **Data transmission** : Collected data is sent via wireless networks such as Wi-Fi, Bluetooth, LoRaWAN or Zigbee.

---

[234] Sundmaeker, et al 2010;

[235] Gubbi, J., et al. (2013). "Internet of Things (IoT): A Vision, Architectural Elements, and Future Directions." *Future Generation Computer Systems, 29*(7), 1645-1660 .

- **Servers and processing**: Data is then stored and analyzed on servers in the cloud, where artificial intelligence algorithms can be applied to extract information and trends (Miorandi et al., 2012) .[236]

IoT sensors use a variety of protocols to transmit data. The most common include MQTT (Message Queuing Telemetry Transport), CoAP (Constrained Application Protocol) and HTTP/HTTPS. Each protocol has its advantages and disadvantages depending on the situation of use and the forensic requirements of transmission (Bertino and Islam, 2017)[237] **. They** have a wide range of applications in several sectors, making them indispensable in the digital age and enabling monitoring of environmental conditions and crop health. For example, soil sensors measure moisture, temperature and nutrient levels, enabling farmers to make informed decisions on irrigation and fertilization. This leads to more sustainable and efficient agriculture, increasing yields while reducing costs (Wolfert et al., 2017)[238] . In smart cities, IoT sensors are used for traffic control, energy management and air quality monitoring. For example, traffic sensors can collect real-time data on traffic flow, enabling better regulation of traffic lights and a reduction in traffic jams. Similarly, pollution

---

[236] Miorandi, D., et al. (2012). "Internet of Things: Vision, Applications and Research Challenges." *Ad Hoc Networks, 10*(7), 1497-1516.

[237] Bertino, E., & Islam, N. (2017). "Botnets and Internet of Things Security." *Computer, 50*(2), 76-79 .

[238] Id.

sensors help authorities monitor air quality and take action when pollution levels exceed acceptable thresholds (Zanella et al., 2014) .[239]

IoT sensors play a crucial role in personalization and home automation. Motion sensors can activate lighting and heating based on the presence of occupants, while temperature and humidity sensors can optimize energy consumption. In addition, these devices can contribute to security, by detecting potential intrusions and fires (Kumpulainen et al., 2020)[240] **. They** are also used in the healthcare sector for patient monitoring and telemedicine. Wearable devices equipped with biometric sensors can monitor vital parameters such as heart rate, oxygen saturation and physical activity. This data can be transmitted to healthcare professionals, facilitating constant medical monitoring and enabling early intervention in the event of problems (Bashar et al., 2020) .[241]

IoT sensors enable rapid decision-making by providing instant data on various aspects of the environment; Automating procedures with sensors can reduce wasted resources and improve efficiency. In areas such as healthcare and smart cities, IoT sensors are helping to improve citizens' quality of life . [242]

They have a number of disadvantages:

---

[239] Zanella, A., et al. (2014). "Internet of Things for Smart Cities." *IEEE Internet of Things Journal, 1*(1), 22-32.

[240] Kumpulainen, S., et al. (2020). "Smart Home Energy Management System Using IoT." *Journal of Energy Storage, 28*, 101424.

[241] Id.

[242] Id.

- **Security issues**: IoT sensors can be vulnerable to cyberattacks, raising concerns about data security and user privacy (Bertino and Islam, 2017).
- **Interconnectivity**: Efficient sensor operation depends on optimal interconnectivity, and any failure in the network can lead to data loss or poor performance (Gubbi et al., 2013) .[243]
- **Initial costs**: Although IoT sensors can reduce costs in the long term, the initial investment in hardware, infrastructure and training can be high, especially for small businesses (Wolfert et al., 2017).[244]

The future of IoT sensors looks bright with the emergence of new technologies and a growing demand for data intelligence. The integration of artificial intelligence and machine learning will enable predictive analytics on a vast array of data, paving the way for even more sophisticated automation and decision-making .[245]

---

[243] Gubbi, J., et al. (2013). "Internet of Things (IoT): A Vision, Architectural Elements, and Future Directions." *Future Generation Computer Systems, 29*(7), 1645-1660.

[244] Id.

[245] Id.

Figure12 : IoT sensors hold promise for emerging agriculture

In addition, advances in communications networks, such as 5G, will improve the speed and capacity of data transmissions, making IoT sensors even more efficient. Efforts to improve security protocols within IoT networks will also intensify, to protect sensitive data and increase user confidence. IoT sensors represent a major advance in data collection and analysis in a variety of sectors. Their ability to provide real-time information transforms our interaction with the environment and contributes to informed decision-making. Although challenges remain in terms of security and inter-connectivity, the benefits offered by these technologies make them a key element of the digital future. As the IoT continues to evolve, sensors will become increasingly essential to applications that improve quality of life, optimize resources and support sustainable business models. As for Farm Management Systems (FMIS) and Analysis software, they are essential for exploiting the data collected by sensors and other technologies. These systems enable farmers to centralize information, perform in-depth analyses and make informed decisions. FMIS offer functionalities such as crop planning, inventory tracking, financial management and yield analysis. With graphical data visualization, farmers can identify trends and adjust their practices accordingly (Khoshnevisan et al., 2019)[246] . For example, farmers can use predictive models to anticipate climate impacts on production and adapt their strategy accordingly.

In an increasingly technology-driven world, the agricultural sector is no exception to the digital transformation. Farm Management Information Systems (FMIS) play a crucial role in this evolution, providing farmers with tools to plan, monitor and evaluate all activities on their farm. These systems facilitate data collection, performance analysis and resource optimization, leading to more sustainable and productive farming. An FMIS is a set of software and IT tools designed to help farmers manage and optimize their farms. They integrate various functions, such as crop management, performance monitoring, financial management, planning and reporting. By making information accessible and usable, these systems help farmers make informed decisions based on hard data (Bramas et al., 2018) .[247]

---

[246] Koshenevisan et al 2019.

[247] Bramas, M., et al. (2018). "Farm Management Information Systems: A Review of Concepts and Applications." *Agricultural Systems, 162*, 61-76.

FMIS can be classified into different categories according to their specific features and use. These catcgories include:

- **Crop management systems**: Focus on cultivation techniques, yield monitoring and input management.
- **Resource management systems**: Track the use of resources such as water, fertilizers and pesticides.
- **Financial management systems**: Help with budgeting, accounting and cost analysis (Klerkx et al., 2019) .[248]

One of the key features of FMIS is their ability to collect and store data from a variety of sources, including IoT sensors and drones. This data can include information on climate, soil, seeds, fertilizers, and much more. Effective management of this data is crucial to informed planning and decision-making (Thompson et al., 2018) . [249]FMIS enables advanced analyses to be performed on the data collected. This includes visualization tools and dashboards that present operational performance from different angles, making it easier to identify trends and anomalies. Predictive analytics also make it possible to anticipate future impacts on production and adjust strategies accordingly (López et al., 2019)[250] . They also provide planning tools that help farmers develop crop calendars, manage resources and plan day-to-day

---

[248] Klerkx, L., et al. (2019). "The Role of Digital Technologies in the Transformation of Agriculture." *Journal of Agriculture and Environmental Ethics, 32*(5), 767-784.

[249] Thompson, D. L., et al. (2018). "The Impact of Precision Agriculture on Farm Operations." *Journal of Management Studies, 55*(4), 721-747.

[250] López, L. C., et al. (2019). "Precision Agriculture and Farm Management: Advancements and Future Directions." *Sustainability, 11*(18), 5002.

activities. This can include yield forecasts based on historical data and climate models. Real-time monitoring of farm performance enables farmers to monitor the health of their crops, optimize resource use and improve productivity. The reports generated by FMIS also help to assess the effectiveness of the various farming practices adopted[251] . FMIS offers many benefits to farmers, contributing to more sustainable and efficient farm management.

By facilitating decision-making based on accurate data, FMIS improve crop productivity. Farmers can optimize seed timing, fertilizer application and irrigation management, resulting in higher yields; enable more efficient resource management, which can reduce operating costs. For example, optimal use of fertilizers and water prevents wastage and surpluses, leading to lower expenses (Thompson et al., 2018)[252] . By enabling precision farming, FMIS contribute to more sustainable agricultural practices. Reducing the use of chemical inputs, precisely managing irrigation and conserving natural resources are all promoted by these systems (Bramas et al., 2018); help farmers identify and manage risks, whether related to weather conditions, crop disease or price fluctuations. Predictive analytics can assess the potential impacts of these risks on production, facilitating preparedness and appropriate response[253] . One of the main challenges facing the adoption of FMIS is the reluctance of some farmers to adopt new technologies. A preference for traditional methods can hinder the

---

[251] Idem.

[252] Id.

[253] Ibid.

implementation of more advanced systems. In addition, lack of training and digital skills can also limit the optimal use of FMIS (López et al., 2019)[254] . The future of FMIS looks bright, with technological innovations and improvements continuing to transform the agricultural sector. The integration of artificial intelligence and Internet of Things technologies into FMIS will enable advanced automation and even more sophisticated predictive analytics. IoT sensors will provide real-time data, further enriching farm performance analysis.

With the falling cost of digital technologies and increasing connectivity, more and more farmers, including those in developing countries, should be able to access FMIS. This could lead to a significant modernization of farming practices worldwide, improving yields and sustainability. FMIS will play a key role in the transition to sustainable agricultural practices. With the growing emphasis on environmental sustainability and pressure to reduce agriculture's carbon footprint, FMIS can offer solutions for an environmentally-friendly agricultural future. Farm Management and Analysis Software (FMIS) represents a significant advance in modern farm management, offering valuable tools for performance analysis, planning and resource optimization. Their benefits, such as improved productivity, reduced costs and more sustainable management, make them essential allies for 21st century farmers.

[254] FMIS ;Lopez et al, 2019.

However, challenges remain, including technological adoption and data security. As innovations continue to emerge, FMIS is set to play an even more central role in transforming the agricultural sector towards more sustainable and productive farming. Precision irrigation is a key component of precision agriculture, aiming to optimize water use while ensuring crop health. Modern irrigation technologies integrate soil sensors and automated systems to adjust irrigation in real time according to plant needs. These systems use sensors to measure soil moisture and trigger irrigation when necessary. This reduces water wastage and increases irrigation efficiency (Schwarz et al., 2015)[255] . Drip irrigation is a technology that enables targeted application of water directly to plant roots. By combining this method with moisture sensors, farmers can optimally manage water resources, preventing water stress and improving crop yields (Kumar et al., 2018).[256]

Irrigation is an essential part of modern agriculture, particularly in arid and semi-arid regions. It maximizes crop productivity by supplying plants with adequate quantities of water. With increasing global food demand and the challenges posed by climate change, irrigation control technologies have evolved to become more efficient, sustainable and intelligent .[257]

[258]Irrigation control technologies play a crucial role in water resource management. They enable :

[255] Schwarz et al 2015.

[256] Id.

[257] Id.

[258] Automatic irrigation system; www.gestiriego;

- **Optimize water use**: By adjusting irrigation needs according to climatic conditions and soil characteristics, these technologies help reduce water wastage.
- **Improve crop yields**: Better irrigation control can lead to more abundant, higher-quality harvests.
- **Reduce costs**: The automation and precision of irrigation systems can reduce the time and resources required for irrigation.
- **Promoting sustainability**: By enabling more efficient irrigation, these technologies help preserve water resources.

[259]Types of Irrigation Control Technology include automated irrigation systems, sensors and monitoring technology, and irrigation management systems. Automated irrigation systems, such as drip and sprinkler, are designed to apply water directly to plant roots or the soil surface, minimizing evaporation and runoff. These systems can be controlled by timers and moisture sensors that regulate irrigation according to the actual needs of the crop. [260]Soil moisture sensors, weather sensors and other monitoring devices play a crucial role in controlling irrigation systems. These sensors provide real-time data on soil moisture, climatic data and crop condition. In this way, farmers can make informed decisions about when and how much water to apply. Irrigation management systems integrate advanced software and data models for optimal irrigation control. These systems enable farmers to schedule irrigation according to weather forecasts, crop

[259] www.rg2i.com

[260] Op.cit.

water requirements and water availability. [261]Adopting advanced irrigation control technologies offers many advantages. Precision irrigation allows direct targeting of areas requiring water, reducing wastage. According to a study by the Food and Agriculture Organization of the United Nations (FAO), the application of high-tech irrigation systems can reduce water consumption by between 30% and 70% compared with traditional techniques (FAO, 2020). Irrigation control technologies enable optimum management of water resources, resulting in higher yields. Research shows that modernized irrigation systems can increase yields of some crops by up to 50% (Smith et al., 2021)[262] . By promoting smarter water use, these technologies contribute to environmental sustainability. Improved irrigation management can reduce problems of soil salinity and groundwater pollution.

Despite the many advantages of irrigation control technologies, there are challenges to their adoption. The initial cost of automatic systems and sensors can be prohibitive for some farmers, particularly in developing countries. In addition, lack of training and awareness of these technologies limits their use. To overcome these challenges, it is essential to promote training and awareness programs. Government initiatives and public-private partnerships can also facilitate access to these technologies. What's more, ongoing innovation in technology is helping to develop more affordable and accessible solutions.

---

[261] Id.

[262] Smith et al 2021.

[263]Irrigation control technologies represent a crucial step towards more sustainable and efficient agriculture. By optimizing water use, improving yields and promoting responsible resource management, they play a vital role in tackling the world's food challenges. However, to realize their full potential, it is essential to overcome the obstacles to their adoption and to continue innovating in this field.

Remote sensing is a technology that uses remote sensors to monitor changes in the agricultural environment. These sensors can be satellite, airborne or even drone-mounted, and provide data on parameters such as temperature, humidity, chlorophyll and biomass. Remote sensing data can be used to create crop variability maps, identify areas requiring specific interventions, and assess the impacts of agricultural practices on the environment (Friedrich et al., 2019)[264]. Remote sensing thus helps farmers adapt their practices to maximize profitability while minimizing environmental impacts. [265]Remote sensing refers to all techniques for obtaining information about a distant object or surface without direct contact. This technology uses data acquisition systems based on various platforms, such as satellites, drones or aircraft. Remote sensing makes it possible to gather information on environmental, climatic and urban phenomena, and to carry out complex analyses in a wide range of fields, from agriculture to natural resource management.

---

[263] Idem.

[264] Friedrich et al 2019.

[265]Idem.

Remote sensing is based on the capture of electromagnetic energy reflected or emitted by the earth's surface. There are two main categories:

- **Passive remote sensing**: This method uses natural energy (such as sunlight). Sensors measure the energy reflected by surfaces. Earth observation satellites such as Landsat or MODIS use this approach.

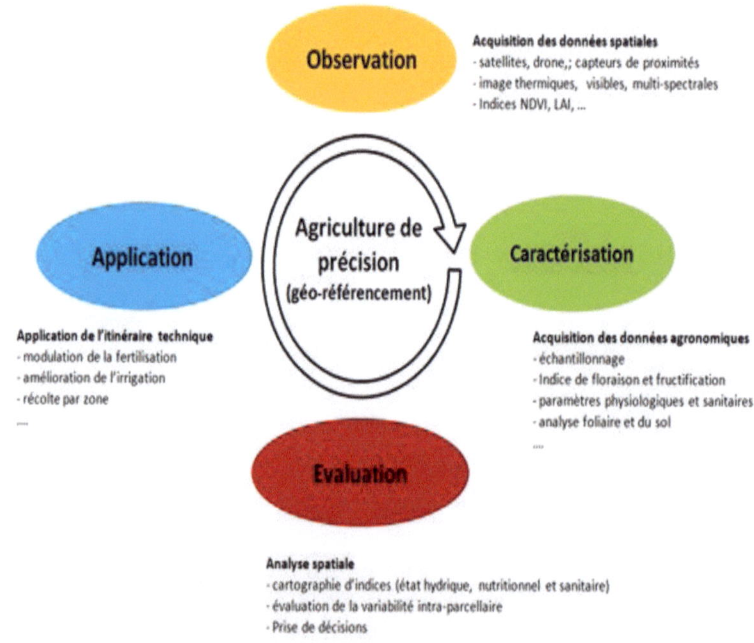

Figure13 : Crop accuracy and geo-referencing

- **Active** remote sensing: In this case, sensors emit their own signal, such as radar or lidar (Light Detection and Ranging). These systems measure the time it takes for the signal to return after hitting an object, providing information about the surface and its structure (Mallet et al., 2015).

Remote sensing systems generally consist of the following components:

- **Platforms**: This includes satellites, aircraft, drones and even ground-based telescopes.
- **Sensors**: Devices that collect data, using different spectral bands from ultraviolet to microwave.
- **Data acquisition and processing systems**: Software and hardware that analyze and interpret collected data to provide useful information.

Remote sensing plays a key role in precision agriculture. It can be used to monitor crop health, determine water and nutrient requirements, and identify diseases. Using multi-spectral images, farmers can assess crop biomass and carry out targeted interventions to optimize yields (Frazier & Page, 2000)[266] . In the field of natural resource management, remote sensing is used to monitor forests, waterways and ecosystems. For example, it can be used to assess the impact of deforestation, monitor water quality and analyze changes in land use (Turner et al., 2015)[267] . Remote sensing data are essential for monitoring and studying climate change. They make it possible to

---

[266] Frazier, P. S., & Page, K. J. (2000) *Water and nitrogen management in precision agriculture using remote sensing*. Journal of the American Society of Agronomy, 92(1), 8-19.

[267] Turner, W., et al. (2015). *Free and open-access satellite data are key to biodiversity conservation*. Nature Ecology & Evolution, 1(1), 15-20.

measure phenomena such as rising sea levels, rising temperatures and changes in glaciers. Satellites such as the European Space Agency's Sentinel system provide valuable data for these analyses (European Space Agency, 2020) .[268]

In the field of urban planning, remote sensing is used to monitor the expansion of urban areas, analyze infrastructures and assess the impact of development projects. It also provides information on urban temperatures to help combat the effects of urban heat islands. One of the main limitations of remote sensing systems lies in the precision and resolution of the data collected. Depending on the study objectives, a suitable spatial, temporal and spectral resolution is required. For example, satellite images can have a resolution ranging from a few meters to several kilometers, which impacts analysis capacity (Jensen, 2007)[269] . The high cost of remote sensing technologies, particularly for space platforms and high-resolution sensors, can act as a brake on their adoption, especially in developing countries. In addition, access to satellite data may be restricted due to political or financial considerations (Huang et al., 2021). The growing use of remote sensing also raises privacy and ethical concerns. For example, monitoring crop fields or urban infrastructure can raise

[268] European Space Agency (2020). *Sentinel Missions*. Retrieved from https://scihub.copernicus.eu/

[269] Jensen, J. R. (2007). *Remote Sensing of the Environment: An Earth Resource Perspective*. Prentice Hall.

questions about data protection and image rights (Clare et al., 2018)[270]. The future of remote sensing is promising thanks to technological advances such as artificial intelligence and big data. The integration of these technologies makes it possible to analyze massive volumes of data to extract accurate, actionable information. For example, machine learning can improve predictive models in crop management . [271]

Remote sensing can play an essential role in promoting sustainable development. By providing crucial data on natural resources and the environment, it can help governments and organizations make informed decisions, monitor the progress of environmental initiatives and anticipate the effects of climate change.

Global initiatives, such as the European Union's Copernicus program, aim to strengthen access to remote sensing data and promote collaborative research. These programs contribute to greater awareness of environmental issues and to solutions based on robust data (European Commission, 2021)[272] . Remote sensing is an indispensable tool for understanding and managing our environment. Thanks to its varied applications, from agriculture to natural resource monitoring, it offers promising prospects for tackling contemporary

[270] Clare, A. T., Maier, H. R., & O'Rourke, S. M. (2018). *Ethical considerations for the use of satellite remote sensing in land management*. Land Use Policy, 76, 559-566.

[271] Khaki, K., et al. (2020). *An Intelligent System for Crop Type Mapping using Hybrid Remote Sensing Data and Machine Learning Techniques*. Computers and Electronics in Agriculture, 168, 105166.

[272] European Commission. (2021). *Copernicus: Europe's Eyes on Earth*. Retrieved from https://copernicus.eu/

challenges such as climate change and sustainable resource management. However, it is essential to overcome the challenges of cost, accuracy and ethics to maximize its potential and ensure fair and sustainable use.

[273]Finally, the key technologies used in precision agriculture play an essential role in improving the productivity and sustainability of farming operations. Their ability to collect, analyze and apply accurate data enables farmers to make informed decisions, optimize resource use and reduce environmental impacts. As the world faces challenges such as climate change and population growth, precision agriculture represents a viable solution for sustainable food production. In the future, the integration of these technologies will continue to transform the agricultural sector, promoting more responsible and efficient resource management. Precision agriculture encompasses numerous technologies, tools and practices such as dose modulation. This practice is carried out in several stages and requires a number of tools, such as satellite images or drones, to collect and process data in order to measure intra-plot heterogeneity

[273] Wang, J., et al. (2017). Soil Sensor Technologies for Precision Agriculture: A Review. *Computers and Electronics in Agriculture, 141*, 259-266.

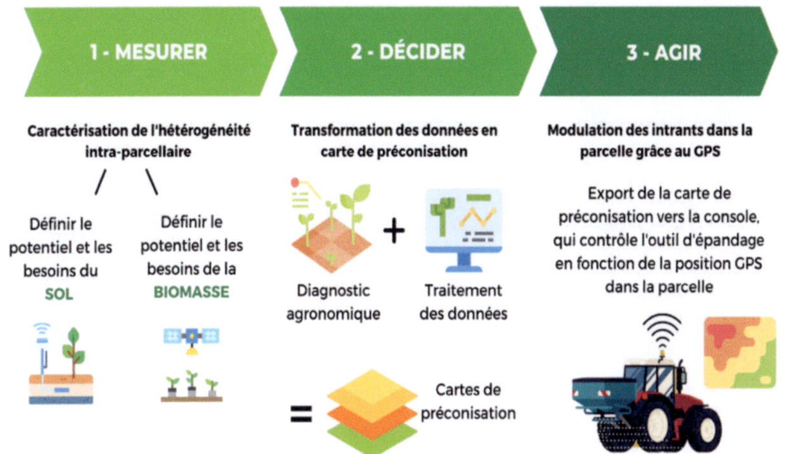

Figure14 : The three (3) stages of precision

2-2- Overview of precision agriculture in Africa

Precision farming is an innovative approach that uses advanced technologies to optimize agricultural yields while reducing costs and minimizing environmental impact. In Africa, this sector is evolving rapidly, offering considerable opportunities for farmers.

However, it faces many challenges, including limited access to technology, inadequate infrastructure and lack of training.

Precision agriculture, of course, relies on technologies such as geographic information systems (GIS), sensors, drones and data analysis. In Africa, food security and increasing agricultural productivity are major challenges.[274] According to the FAO (Food and Agriculture Organization of the United Nations), Africa needs to

[274] FAO (2021).

double its food production by 2050 to feed a rapidly growing population (FAO, 2021). Precision agriculture can play a key role in achieving this goal. [275]Precision farming technologies are varied and constantly evolving. Among the most widely used in Africa are Drones and remote sensing (Dandois and Ellis, 2020)[276] ; Geographic Information Systems (GIS); Sensors and IoT.

Several initiatives have been launched to promote precision agriculture on the continent:

- **The Precision Agriculture for Development (PAD) project**: launched in 2015, this project aims to use mobile technologies to provide farmers with precise recommendations. It has been implemented in several African countries, including Kenya and Uganda.[277]

- **The "Agri-Tech East" program**: Initiated in Zimbabwe, this program aims to integrate digital technologies into agriculture to improve production and sustainability.[278]

-**Start-ups**: Many start-ups, such as Twiga Foods (Kenya) and Zambian Agribusiness (Zambia), are adopting models based on precision agriculture to help farmers access markets while optimizing their yields.

---

[275]Bălteanu, D., et al. (2021). "GIS Applications in Agriculture: Evidence from Ghana." Romanian Journal of Geography, 65(1), 45-60.

[276] Dandois and Elias 2020.

[277] Padmanaban, P., et al. (2021). "Precision Agriculture for Development: Evaluating the Impact of Mobile Technology on Smallholder Farmers." World Development, 138, 105233.

[278] Mavhunga, I., et al. (2022). "Agri-Tech East: An Approach to Implement Precision Agriculture in Zimbabwe." African Journal of Agricultural Research, 17(3), 120-130.

Although precision farming offers considerable advantages, several challenges remain: Many farmers in Africa, particularly those in the informal sector, do not have access to precision technologies due to high costs or lack of infrastructure. The adoption of technologies requires adequate training. Unfortunately, lack of education and awareness of these tools limits their use. Rural infrastructures in Africa are often inadequate, complicating the implementation of precision technologies. Roads, Internet access and electricity remain major obstacles (FAO, 2021)[279] . Agriculture in the Sahelo-Sudanian zones in particular is practiced under difficult agroecological conditions characterized by great variability in climatic parameters (low and unpredictable rainfall, high temperatures), low soil fertility and attacks by pests such as insects, diseases and weeds. Farmers are also faced with capital, labor and land pressure constraints. Public support services are under-resourced, hampering producers' access to extension services and credit. The random nature of markets makes it difficult to plan agricultural activities. In these conditions, it is important to develop farming practices for producers that enable efficient/effective use of resources and risk reduction. In this document, the term Sahelo-Sudanian zone refers to the Sahelian zone and the Sahelo-Sudanian zone of West Africa, which respectively receive 250-500 mm and 500-900 mm of rain per year (FAO, 2006).[280]

[279] Id.

[280] FAO (2006).

Precision Agriculture is an approach to improving the efficient use of resources that has hitherto been practised only on commercial farms. The use of weather forecasting, Global Positioning System (GPS), Geographic Information System (GIS), remote sensing and advanced farm mechanization is making Precision Farming increasingly attractive to commercial farmers (Tiwari and Jaga, 2012)[281] . The efficient use of resources is an important factor in Precision Agriculture in the Sudano-Sahelian zone, a more simplified version of which is the subject of this paper. The benefits of Precision Agriculture include more efficient farming operations.

The benefits of Precision Agriculture include faster farming operations (sowing, pesticide application, weeding and harvesting), fertilizer and irrigation application that takes into account soil properties, rainfall and crop protection techniques[282] . Farmers in the Sudano-Sahelian zone are already using elements of Precision Agriculture such as high-yielding varieties, good soil and mulching on degraded soils (Lamers et al., 1998)[283] . However, the concept of Precision Agriculture is rarely used in the context of rainfed agriculture. There is a fairly sound basic knowledge of the agro-ecological constraints affecting agriculture in the Sudano-Sahelian zone, and technologies to overcome these constraints have been developed. These technologies include improved varieties, seed

[281] Tiwari and Jaga, 2012.

[282] Id.

[283] Lamer et al 1998.

soaking, microdosing, manure spreading, mulching, tree introduction and production. (Schlecht et al., 2006; Aune and Bationo, 2008). [284]

However, despite this wealth of knowledge and the availability of technologies, adoption of improved agronomic practices has been slow. There appears to be a challenge in developing technologies that are suitable for the existing farming system. Technologies such as mulching and tree planting are particularly difficult to implement because of the free grazing system that prevails in the Sahelo-Sudan zone after crop harvesting. The way crops are grown in the Sahelo-Sudanese zone makes Precision Farming a feasible option. Sorghum and millet, the main crops in the zone, are sown at low densities ranging from 5,000 bunches per hectare in marginal Sahelian conditions to 25,000 bunches per hectare in more favorable conditions in the Sudanian zone (Memento de l'agronome, 2009). These low sowing densities mean that the input can be applied only to the poquet. This creates a microenvironment around the plant where there is a greater concentration of nutrients and water. Resources and inputs are scarce in the Sahelian and Sudanian zones, so it's a sensible agronomic technique to concentrate resources/inputs close to the plant rather than spreading them across the whole field. This practice favors the effects of seed treatment, the use of organic and mineral fertilizers, the use of fungicides/insecticides and mechanization. Precision Agriculture in the Sahelo-Sudanian regions should not only increase yield and income for producers, but also make agriculture less vulnerable to

---

[284] Schlecht et al 2006; Aune and Bationo 2008.

risks and reduce workload. This paper reviews existing studies on crop and soil management in the drylands of West Africa and synthesizes the results on the use of Precision Agriculture principles.

[285] The study assesses how different technologies can be combined to increase land productivity and reduce labor. Farming in the Sahel-Sudan region is practiced under difficult agroecological conditions characterized by high variability of climatic parameters (insufficient, unpredictable rainfall and high temperatures), low soil fertility, pest attack, disease and weed infestation. Farmers also face constraints in terms of funds (income), labor and land. Public services are under-resourced, limiting 69 Precision agriculture: an option for improving land and labor productivity in the Sudano-Sahelian zone of West Africa farmers' access to extension services, infrastructure and credit. Markets are unpredictable, making it difficult for farmers to plan their agricultural activities. In such conditions, it is important to develop farming practices that make efficient use of resources and reduce farm risks. (Aune Jens Bernt, Coulibaly Adama and Kamkam Woumou l'agriculture de précision : une option d'amelioration de la productivité des terres et du travail en zone soudano-sahelienne de l'Afrique de l'Ouest. [286]The future of precision agriculture in Africa looks promising, provided existing challenges are overcome. Governments and private investors need to work together to improve access to rural

---

[285] Id. Aune Jens Bernt, Coulibaly Adama and Kamkam Woumou; l'agriculture de precision: une option d'amélioration de la productivité des terres et du travail en zone soudano-sahelienne de l'Afrique de l'Ouest.

[286] Idem.

infrastructure. Specific training initiatives are needed to make farmers aware of the benefits of precision technologies. Collaboration between the public and private sectors can foster the development and implementation of precision farming technologies. It is crucial to develop financing models adapted to small-scale farmers so that they can invest in these technologies.

By overcoming these obstacles, Africa can significantly improve its agricultural production, guarantee food security and foster sustainable development. A strong commitment from governments, businesses and farmers is essential to realize the potential of precision agriculture on the continent .[287]

2-2-1- Adoption of precision farming technologies

[288]The adoption of precision farming technologies by farmers is essential to meet the contemporary challenges facing agriculture, particularly in the face of population growth, climate change and the need to guarantee food security.

Throughout the history of agriculture, new tools, technologies and innovations have led to ever more productive growing seasons, marked by a particularly spectacular increase in the second half of the twentieth century. While this Green Revolution was seen as an improvement for farmers, the resulting shift in farmland management

---

[287] Nkonya, E., et al. (2020). "Understanding the Adoption of Precision Agriculture in West Africa: Insights from Farmers." Agricultural Economics, 51(5), 741-756.

[288] Idem.

towards greater dependence on external inputs has led almost half of American fruit and vegetable growers to operate at a loss. Worse still, the resulting dependence on fossil fuels and agro-petrochemicals has only led to soil degradation, collapsing pollinator populations, worsening air quality and diminishing farm sustainability.

[289]It's understandable that, under these conditions, many farmers are unable or reluctant to adopt new technologies. Yet many of today's advances in AgTech are designed to help farmers implement efficient, productive strategies that protect their livelihoods in the long term. There's always a risk when it comes to adopting a new technology, but by carefully weighing up the options and potential gains, it may be worth taking a chance.

Over the years, agriculture and technology have continued to evolve together. Of course, the term "technology" refers not only to specific tools, but also to :

- Agricultural innovations and practices, such as understanding the concept of soil exhaustion
- Understanding ecology and preservation procedures
- Scientific progress

[290]Technology is at the forefront of modern agriculture and plays a key role in improving efficiency, sustainability and productivity. Precision farming and digital agriculture technologies, such as GPS-connected tractors, drone monitoring and soil sensors, enable farmers to manage

---

[289]Precision agriculture; leshorizons.net

[290] Ibid.

their crops and livestock with greater precision and efficiency. Farmers can monitor field conditions in real time, enabling them to make decisions and intervene at the right time.

[291]Automated machines help farmers to manage considerable workloads and reduce labor shortage problems. In addition, artificial intelligence applications in the field of predictive analytics help anticipate yield results, potential pest infestations and weather-related events, thus contributing to risk mitigation. Mobile applications and online platforms provide real-time market data, helping farmers to make more informed sales decisions. Electrification eliminates emissions for cleaner, safer working environments, while saving a farm thousands of dollars each year in diesel and maintenance costs. [292]Precision farming tools and agricultural equipment, which enable real-time control and monitoring of farming operations. Instead of manually checking every nook and cranny of the farm, farmers can monitor operations using drones and IoT devices, short for "internet of things". These tools provide real-time data on crop health, soil conditions and weather, enabling farmers to make timely adjustments and optimize yields.

[293]To effectively adopt new agricultural technologies, it's useful to follow a basic outline. The first step towards adopting new agricultural technologies is to identify your farm's needs and objectives. Do you want to increase crop yields? Would you like to

---

[291] Id.

[292] www.cairns.info

[293] Precision agriculture;

reduce water consumption or improve soil health? Perhaps you want to be less dependent on external inputs. By identifying these needs and objectives, you can determine the type of technology that would be most beneficial to your operation. Once you've identified your needs and goals, it's time to dive into the world of AgTech and explore the technologies and innovations available. This can range from precision farming tools, automated machinery and drone technology to software solutions for farm management and data analysis. Look for technologies that match your objectives and contact manufacturers for more information or demonstrations.

[294]Once you've done your research, draw up a short list of technologies that could be useful for your operation, and evaluate these solutions in a little more depth. Consider key factors such as cost, ease of implementation, compatibility with your existing systems and potential return on investment. [295]Consider talking to other farmers who have adopted these technologies or to get an idea of how they perform in the real world.

So once you've selected the technology that best suits your needs, it's time to implement it. This may involve purchasing equipment, installing software and training staff. Be prepared for an adaptation period during which you and your team will get used to the new technology. Once the technology is in use, be sure to evaluate its effectiveness on an ongoing basis. Is it helping you achieve your

---

[294] Op.cit.

[295] Id.

objectives? Be sure to communicate openly with the manufacturer to provide feedback and discuss improvements that could help your operations. Adopting new technology is a dynamic process that requires constant evaluation and adaptation. Technology in agriculture is creating a landscape that embraces sustainable practices and increases resilience while meeting the world's growing demand for food.

Future prospects for the adoption of precision farming technologies in Africa and other developing regions are promising. With increased investment in agriculture and innovation, precision technologies may become more accessible. Advances in mobile technologies and cloud-based applications could also facilitate adoption, especially among small-scale farmers (FAO, 2021).[296] [297]The adoption of precision farming technologies is crucial to meeting contemporary agricultural challenges. Although several obstacles exist, appropriate strategies can encourage wider adoption. By investing in training, financial support and infrastructure improvements, it is possible to significantly increase agricultural productivity while minimizing environmental impact. The future of agriculture lies in the ability of the whole sector to integrate these new technologies in an efficient and sustainable way.

[296] FAO (2021).

[297]Padmanaban, P., et al. (2021). "Precision Agriculture for Development: Evaluating the Impact of Mobile Technology on Smallholder Farmers."

2-2-2- Factors facilitating and constraining adoption

The integration of artificial intelligence and robotics in agriculture offers enormous potential for increasing productivity while reducing environmental impact. By leveraging these technologies, modern agriculture can become more precise, efficient and sustainable, meeting the growing needs of a global population.

Unleashing the potential of AI in agriculture can not only improve efficiency and productivity, but also promote more sustainable use of agricultural resources. This can contribute to greater food security and reduced environmental impact. Precision agriculture, which uses advanced technologies to optimize production. Among the factors facilitating adoption are :

The use of agricultural technologies offers considerable potential for improving farm productivity, sustainability and profitability. However, adoption of these technologies varies widely from region to region and from farmer to farmer. Understanding the factors that facilitate or hinder this adoption is essential to maximizing the benefits.

-Access to Technology and Infrastructure

The availability and accessibility of precision technologies are essential prerequisites for their adoption. Technological advances such as drones, IoT sensors and farm management software must be accessible at a reasonable price. Studies show that initiatives that promote access to these technologies, for example through subsidies or co-financing, can significantly increase their adoption (Khan et al., 2019) .[298]

-Education & Training

[299]Training farmers in the use of new technologies is a key factor. When farmers are trained and educated on the benefits and use of precision tools, they are more likely to adopt them (Nkonya et al., 2020)[300] . Training programs through NGOs and government agencies have shown positive results in several developing countries.

-Perceived Benefits of Technology

The tangible benefits associated with the use of precision farming, such as increased yields, water savings and reduced input costs, play a crucial role. According to a study conducted by Padma[301] , farmers who understand these benefits are much more likely to adopt precision technologies.

-Institutional Support

The role of governments and agricultural institutions is also critical. The development of policies favorable to precision technologies, as well as financial and logistical support, can encourage farmers to switch to these new methods. For example, national agricultural

---

[298] Khan, A. R., et al. (2019). "Barriers to Adoption of Precision Agriculture in Rural Africa: A Systematic Review."

[299] Idem.

[300] Id.

[301] Padmanaban, P., et al. (2021). "Precision Agriculture for Development: Evaluating the Impact of Mobile Technology on Smallholder Farmers.

development programs incorporating precision technologies are showing promising results in several countries.

In addition, with regard to constraining factors, we note :

-High initial costs

The costs of acquiring and installing precision technologies can be prohibitive for many farmers, particularly those managing smallholdings. According to Khan et al. (2019)[302] , the investments required to implement precision irrigation systems or remote sensing devices can limit access to these technologies.

-Technological complexity

The complexity of precision technology systems can also pose a challenge. Many farmers, especially those who lack technical training or prior experience with the technology, may feel overwhelmed and reluctant to invest in these tools (Mavhunga et al., 2022) .[303]

-Poor awareness and incomplete information

Lack of awareness of available technologies and potential results is a major obstacle. Farmers may be unaware of innovations that could improve their yields, limiting their interest in precision technologies.[304]

---

[302] Id.

[303] Mavhunga, I., et al. (2022). "Agri-Tech Innovations in Zimbabwe: Building Precision Agriculture."

[304] Id.

-Inadequate infrastructure

In many regions, particularly in developing countries, inadequate communications and transport infrastructure can hamper the adoption of precision technologies. Lack of adequate Internet connectivity limits the use of cloud-based solutions and mobile applications.

The adoption of precision farming technologies is influenced by a multitude of facilitating and constraining factors. Measures to increase access to technologies, provide adequate training, strengthen institutional support, and raise farmer awareness are essential to encourage this adoption. At the same time, initial costs, complexity, lack of awareness and insufficient infrastructure represent significant challenges that need to be overcome. By addressing these constraints strategically, it is possible to maximize the benefits of precision farming, both for farmers and for society as a whole.

2-3-Impact of precision farming on crop yields

Precision farming is an innovative approach based on the use of advanced technologies to optimize agricultural yields while reducing environmental impact. The rise of tools such as drones, sensors, geographic information systems (GIS) and data analysis has revolutionized agricultural practices. In a global context marked by a growing population and the challenges of climate change, it is imperative to explore how these technologies can transform food production. Precision agriculture is defined as an approach to farm management that uses geo-spatial data and digital technologies to observe, measure and react to field variables. This model is based on the idea that each plot of land has unique needs, and that the application of resources such as water, fertilizers and pesticides must be adjusted to maximize efficiency and yield (Zhang et al., 2016)[305] . Precision techniques include the use of sensors for soil monitoring, drones for crop surveillance, and data analysis software for informed decision-making.

Indeed, yields are increasing thanks to precision technologies: the use of data and predictive analysis; precision irrigation; nutrient management; pest and disease reduction. Precision farming technologies enable the collection of real-time data on various crop parameters, such as soil moisture, temperature and plant health. For example, the use of soil sensors can help farmers determine the optimal time to irrigate (Pérez-Ruiz et al., 2020)[306] . Studies have shown that farmers who integrate predictive analytics systems can increase their yields by up to 30% compared to those who don't (Thorp et al., 2018)[307] . Precision irrigation is one of the most promising applications of precision agriculture. By using sensors to monitor soil moisture, farmers can optimize water application. Studies have shown that precision irrigation can reduce water requirements by 20-50% while increasing yields by 10-30% (Hedley et al., 2017)[308] . What's

---

[305] Id.

[306]Perez- Ruiz et al ; 2020.

[307] Thorp et al; 2018.

[308] Hedley, C., et al. (2017). "The Impact of Precision Irrigation on Water Use Efficiency and Crop Yield." *Irrigation Science*, 35(1), 31-44.

more, this approach contributes to the sustainability of water resources, a crucial issue in arid zones. [309]Irrigation is a crucial element of agriculture, particularly in regions where rainfall is irregular or insufficient to support crop growth.

However, traditional irrigation can lead to significant water wastage and adverse environmental impacts. In response to these challenges, the concept of precision irrigation has emerged, offering a more efficient and sustainable approach to water management. Precision irrigation combines advanced technologies to monitor, plan and manage water use in a way that maximizes crop yields while minimizing the resources required.

Precision irrigation refers to the use of modern technologies to provide crops with the optimum amount of water at the right time and in the right place. This method relies on real-time data on soil moisture, climatic conditions and specific crop needs. Key techniques include :

- **Soil sensors**: These devices measure soil moisture, temperature and other parameters to determine irrigation requirements.
- **Automated control systems**: These adjust the water supply according to the data collected, enabling dynamic irrigation management.
- **Drip irrigation systems**: These systems deliver water directly to the plant root zone, limiting water loss through evaporation and runoff.

[309] Idem.

One of the main advantages of precision irrigation is the significant reduction in water consumption. According to several studies, this approach can reduce water use by 30-50% compared to traditional irrigation methods[310] . Farmers can save on water costs while maintaining or even increasing yields. By delivering the optimum amount of water to plants, precision irrigation systems help to improve crop health and, consequently, yields. Research has shown that plantations using precision irrigation can experience yield increases of up to 15-30% (Zhang et al., 2016)[311] . This is particularly important in areas where competition for water is strong. Precision irrigation not only reduces water consumption, but also minimizes negative effects on the environment. By limiting irrigation to areas where it is necessary, nutrient leaching and soil salinization are prevented. This contributes to the health of surrounding ecosystems and the sustainability of water resources (Postel, 2014)[312] . Precision irrigation relies on various technologies to optimize water use, such as remote sensing, soil sensors, irrigation management systems, challenges and limitations. Remote sensing uses sensor-equipped satellites or drones to monitor crops and soil conditions on a large scale. These devices provide valuable information on soil moisture, crop health and water requirements, enabling farmers to make informed decisions (Mulla,

---

[310] Id.

[311] Zhang, Q., et al. (2016). "Precision Agriculture: A New Approach for Sustainable Development." *Journal of Environmental Management*, 182, 338-344.

[312] Postel, S. (2014). "Growing a World of Water: The Challenge of Agriculture." *Water Policy*, 16(S1), 11-26.

2013)[313] . Sensors embedded in the soil provide real-time data on moisture, temperature and conductivity. This enables precise control of irrigation, adjusting water flow according to specific crop needs (Garg et al., 2021)[314] . These automated systems analyze sensor data and adjust irrigation schedules and quantities. For example, specialized software can predict when to irrigate based on weather forecasts and historical soil moisture data.

Although precision irrigation offers many benefits, its adoption faces certain challenges: One of the main barriers to the adoption of precision irrigation is the initial cost of equipment and technologies. Advanced irrigation systems, sensors and control parts can represent a significant investment for farmers, particularly small-scale ones. The effectiveness of precision irrigation depends largely on farmers' ability to use these technologies. It is therefore essential to provide adequate training and technical support to ensure optimal use (Dandois and Ellis, 2020)[315] . In some rural areas, a lack of infrastructure can limit access to the technologies needed for precision irrigation. Real-time data often requires reliable Internet connectivity, which may be lacking in some areas. The potential for precision irrigation is immense, especially as the challenges of climate change, population

[313] Mulla, D. J. (2013). "Twenty Five Years of Remote Sensing in Precision Agriculture: Key Advances and Remaining Challenges." *Precision Agriculture*, 14(1), 2-20.

[314] Garg, S., et al. (2021). "Advancements in Soil Moisture Measurement Technology for Precision Irrigation." *Agricultural Water Management*, 243, 106408.

[315] Dandois, J. P. & Ellis, E. C. (2020). "Drone Applications in Precision Agriculture: A Review." *Journal of Field Robotics*, 37(1), 140-152.

growth and water resource management continue to grow. Technological innovations, such as artificial intelligence systems and the Internet of Things (IoT), promise to make precision irrigation even more efficient and accessible[316] . To encourage the adoption of precision irrigation, it is crucial that governments and institutions put in place incentive policies, such as subsidies for the purchase of equipment or training programs for farmers. Investment in research and development is also essential to further improve precision irrigation technologies and make them accessible to farms of all sizes. [317]Precision irrigation represents a significant step towards sustainable and efficient agriculture. By integrating modern technologies to manage water intelligently, farmers can not only improve their yields, but also contribute to the preservation of natural resources. To realize the full potential of precision irrigation, it is necessary to overcome adoption challenges, encourage training and develop supportive policies. Ultimately, precision irrigation can play a decisive role in global food security and the sustainability of agricultural systems. Nutrient management techniques through precision agriculture enable the targeted application of fertilizers. Initial soil tests and remote sensing-based diagnostics help farmers understand the specific nutritional needs of each plot (Sullivan et al., 2019). This precision in fertilizer application can lead to increased crop yields, with studies indicating increases of up to 25% in some crops (Gutiérrez et al.,

---

[316] Id.

[317] Id.

2020)[318] . The use of drones and multi-spectral cameras to monitor crop health also enables early detection of disease and pest infestations. By acting quickly, farmers can minimize losses. According to a study by Yang et al. (2019)[319] , the use of remote sensing has reduced crop losses by up to 15%, contributing to higher overall yields.

Precision farming has the potential to transform modern agriculture by dramatically improving yields while reducing environmental impact. Despite the challenges that remain, the benefits to agricultural yields justify investment in these technologies. To realize this potential, it is crucial to strengthen education, facilitate access to technologies, and develop favorable policies. With the right support, precision agriculture could play a central role in global food security and the sustainability of natural resources.

2-3-1- Case study in Africa

Precision farming is transforming the global agricultural landscape, and its adoption in Africa is particularly promising. With challenges linked to food security, resource management and climate change, precision agriculture offers an opportunity to optimize yields while minimizing costs and environmental impacts. [320]Africa faces a variety of agricultural constraints, including low yields, growing food

[318] Gutierrez et al; 2020.

[319] A Study by Yang et al, 2019.

[320] Idem.

insecurity and varied climatic conditions. Smallholders represent a majority of farmers in Africa, often with limited access to resources and modern technologies. The introduction of precision agriculture, combined with targeted interventions, could address the need to improve productivity without compromising the sustainability of farming systems.

In Kenya;[321] A study by Mboga et al (2021) explored how the use of drones for crop monitoring improved maize yields among smallholder farmers in Kenya. Drones equipped with multi-spectral sensors were used to assess crop health and identify areas requiring attention (irrigation, fertilization, etc.). The results of the study showed a 24% increase in yields for farmers using this technology, compared with those using traditional practices.

In South Africa, an initiative has been set up to use soil sensors to better manage nutrients in maize and soybean crops.[322] According to a study by Moshabela et al (2020), the use of sensors led to a better understanding of specific crop needs, resulting in a 20% reduction in fertilizer costs and a 15% increase in yields. This approach also helped maintain soil quality and reduce nutrient leaching, which is crucial for long-term sustainability. [323]A study by Chhod et al (2019) in Tanzania showed how precision irrigation has been integrated into rice cultivation. Through the use of soil moisture sensors and drip

---

[321] A Study by Mbonga et al, 2021.

[322] UNE Study by Moshabola M; et al, 2020.

[323] Chhode, S,et al.(2019). Impact of Pfecision irrigation on rice Yield in Tanzania. Journal of irrigation and drainage Engineering, 145(1) , 04018030.

irrigation systems, farmers were able to reduce water consumption by 40% while increasing rice yields by 30%. This project has demonstrated that optimized irrigation techniques can not only improve yields, but also introduce a more sustainable approach to water management.

The case studies above illustrate the use of several precision farming technologies, including :

- **Drones and remote sensing**: used to assess crop health and optimize interventions (Mboga et al., 2021).
- **Soil Sensors:** Help manage plant nutrition and better understand water requirements (Moshabela et al., 2020) .[324]
- **Drip Irrigation Systems:** Reduce water use while maximizing yields (Chhod et al., 2019) .[325]

These technologies enable farmers to make data-driven decisions, leading to more efficient resource management.

In California, where water is a precious resource, the adoption of precision agriculture has transformed irrigation practices. Farmers using soil moisture sensors have managed to reduce water consumption by 30%, while increasing yields by 20% for crops such as grapes and vegetables (Bach et al., 2018). The results demonstrate the effectiveness of these technologies in an arid climate. In parts of sub-Saharan Africa, the use of precision agriculture is proving its worth in improving yields. Projects have been set up to integrate drone

---

[324] Idem.

[325] Id.

technology into farming practices. For example, in Kenya, farmers using drone mapping services have seen a 24% increase in maize yields due to improved resource management (Mboga et al., 2021) .[326]

Despite notable successes, the adoption of precision agriculture in Africa is not without its challenges: Precision technologies can require significant upfront investment, which can be a barrier for smallholders (Khan et al., 2019). Lack of training and skills among farmers often limits the use of advanced technologies (Dandois and Ellis, 2020)[327] . In some rural areas, lack of infrastructure, including electricity and Internet access, can hinder the use of technologies.

For precision farming to have a significant impact on agriculture in Africa, it is essential to invest in research, training and supportive policies. Governments, NGOs and the private sector must work together to create awareness and training programs to help farmers adopt these technologies. In addition, it is crucial to develop affordable and accessible solutions so that smallholders can benefit

[328]Precision agriculture presents an exceptional opportunity to improve agricultural yields in Africa, while addressing the contemporary challenges of sustainability and food security. Case studies from Kenya, South Africa, Tanzania, California and parts of sub-Saharan Africa illustrate the potential of these technologies to transform farming practices. By overcoming the challenges associated

---

326 Id.

327 Dandois, J. P., & Ellis, E. C. (2020). "Drone Applications in Precision Agriculture: A Review." *Journal of Field Robotics*, 37(1), 140-152.

328 Id.

with the adoption of precision agriculture, the African continent can achieve significant gains in productivity, sustainability and resilience to the impacts of climate change.

2-3-2-Comparative analysis of yields before and after adoption

[329]Agriculture, as a fundamental sector of the global economy, has undergone major transformations over the decades. The adoption of modern techniques, from the use of improved seeds to optimized irrigation, has changed agricultural yields in many countries. The history of agriculture shows an evolution in agricultural yields due to several factors. Before the Green Revolution of the 1960s, many developing countries suffered from low yields due to traditional farming methods, lack of access to technology and unfavorable climatic conditions. One of the most convincing examples of the impact of modern techniques on yields is the adoption of improved seeds. In East Africa, for example, studies have shown that the use of modern techniques has led to an increase in yields.

Mechanization and improved irrigation systems are other key factors in increasing agricultural yields. In countries such as Brazil, the use of modern tractors and irrigation technologies has contributed to a substantial increase in yields. For example, soybean yields in Brazil

[329] FAO (2020). "Egypt: Water Management for Agriculture."

have almost doubled, from 2.2 tonnes per hectare in 1990 to 4.0 tonnes per hectare in 2019 (CONAB, 2020) .[330]

In modern Egypt, the introduction of advanced irrigation systems has enabled water to be used more efficiently, increasing crop yields by 20-40% depending on the crop (FAO, 2020)[331] . This development has enabled farmers to cope better with periods of drought. Technology transfer and farmer training play a crucial role in the adoption of modern techniques. Training programs in sub-Saharan Africa have shown that technical support has led to an increase in agricultural yields of 25% on average (World Bank, 2018) .[332]

Despite the successes, several challenges limit the universal adoption of modern techniques. Firstly, the costs associated with the technology can be prohibitive for small-scale farmers, particularly in developing countries. In addition, reliance on chemical fertilizers can lead to environmental sustainability issues, such as soil degradation and pollination of water resources.

In addition, the effects of climate change are making some technologies less effective in certain regions, making it necessary to continually reassess their use (IPCC, 2019)[333] . The resilience of farming systems must become the priority in the future development of agricultural techniques. Comparative analysis of agricultural yields

---

[330] CONAB (2020). "Relatório de Acompanhamento da Safra Brasileira de Grãos."

[331] Id.

[332] World Bank (2018).

[333] IPCC (2019). "Climate Change and Land." Intergovernmental Panel on Climate Change.

before and after the adoption of modern techniques shows a general trend towards higher yields thanks to these innovations. However, to maximize these benefits, it is essential to address the associated challenges and ensure equitable access to the new technologies. Ultimately, the future of agriculture will depend on the ability of countries to integrate sustainability into their agricultural development, while constantly innovating to meet global challenges.[334]

---

[334] Id.

Chapter 3:

The role of artificial intelligence in precision agriculture

Agriculture has undergone radical transformation in recent decades, and one of the most significant developments is the emergence of precision farming. This concept is based on the use of advanced technologies to observe, measure and respond to variations within fields. Artificial intelligence (AI) plays a central role in this dynamic, helping to optimize resources, increase yields and reduce environmental impact. Precision agriculture involves managing crop varieties and soil conditions with great precision to maximize production while minimizing inputs such as water, fertilizers and pesticides. [335]According to the Food and Agriculture Organization of the United Nations (FAO), the main objective is to produce more, with less (FAO, 2020). AI, in this context, is used to process large amounts of data from various sources, such as sensors, drones and satellites, to improve decision-making (Wolfert et al., 2017)[336] . AI enables real-time crop monitoring. Drones equipped with multi-spectral cameras can collect data on plant health, quickly identifying disease or nutrient deficiencies. Machine learning algorithms analyze this data to provide recommendations on which treatments to apply. For example, a study by Zhang et al proved that using AI to analyze crop images could

[335] www.fao.org

[336] Wolfert, S., et al. (2017). "Big Data in Smart Farming - a review." *Agricultural Systems*, 153, 69-80.

reduce yield losses caused by disease by up to 20%. Efficient water management is crucial in agriculture, especially in arid regions. AI, combined with the Internet of Things (IoT), can analyze weather data, soil moisture content and other variables to determine crop irrigation requirements. For example, intelligent systems such as those developed by Israeli company CropX use AI to adapt irrigation in real time, enabling a 30% reduction in water consumption while maintaining yields (CropX, 2021)[337] . AI also helps optimize the use of agricultural inputs, such as fertilizers and pesticides. Using predictive models based on historical and real-time data, farmers can apply the right quantities at the right time. [338]A study by Chapman et al (2019) showed that an AI-based approach to fertilizer application could reduce costs by 15% while improving soil health. AI also makes it possible to predict yields more accurately. By combining historical data on crop cycles with factors such as weather and soil analysis, models can estimate yields for different crops. According to a World Bank report (2020)[339] , this information enables farmers to make informed crop planning and market decisions. One of the main benefits of using AI in precision agriculture is improved resource efficiency. By using accurate data to guide decisions, farmers can reduce waste and optimize their processes.

AI-based technologies have the potential to increase agricultural yields. Intelligent systems can quickly identify problems in crops,

[337] CropsX 2021; The fucture of sustainable irrigation.

[338] A Study by Chapman et al.2019.

[339] World Bank 2020.

enabling early intervention and, consequently, better production. By optimizing the use of inputs, AI helps to reduce the environmental impact of agriculture. Fewer fertilizers and pesticides are applied, reducing pollution and protecting biodiversity.

Despite its benefits, integrating AI into precision agriculture does come with challenges. Among these, the high cost of the technology and the lack of adequate training for farmers are limiting factors. In addition, concerns about data privacy and technology dependency are emerging in this new agricultural landscape. The cost of smart technologies can be prohibitive for small-scale farmers. Substantial initial investments are needed to set up the required infrastructure and acquire the tools needed to apply AI-based systems (Garnett et al., 2013)[340] . The digitization of agriculture requires farmers to possess technical skills that they do not necessarily have. Training and education programs are needed to ensure that farmers can take advantage of these technologies (Klerkx et al., 2019)[341] . The future of precision agriculture will be strongly influenced by the evolution of artificial intelligence. As technologies continue to improve, the integration of more sophisticated systems will enable even greater optimization of agricultural processes. In addition, the growing use of open data could foster wider access to AI technologies, making precision farming accessible to farmers of all levels.

[340] Garnett, T., et al. (2013). "Sustainable Intensification in Agricultural Systems."

[341] Klerkx, L., et al. (2019). "The Role of Digital Technology in the Future of Food and Agriculture."

[342]Artificial intelligence is playing a growing role in precision agriculture, providing innovative solutions to the challenges of global food production. By enabling decision-making based on accurate data, AI facilitates more efficient, sustainable and profitable agriculture. However, for its full potential to be realized, it is essential to overcome barriers related to cost, education and technology integration . The future of agriculture will depend on the ability to combine technological innovation and social inclusion.

Precision agriculture is a revolutionary concept that uses advanced technologies to improve the efficiency, profitability and sustainability of farming practices. Artificial intelligence (AI) is at the heart of this transformation, enabling large-scale data analysis and informed decision-making. By optimizing every stage of agricultural production, from soil preparation to harvesting, AI helps not only to increase yields, but also to reduce environmental impacts.

3-1-1-Presentation of artificial intelligence and its applications in agriculture

[343]Artificial intelligence (AI) refers to the simulation of human intelligence processes by computer systems. This includes learning (e.g. machine learning), reasoning, perception and language comprehension. In other words, AI enables machines to process data, recognize patterns, make decisions and even learn new information

[342] Id.

[343] www.netapp.com

over time. [344]In agriculture, AI has emerged as a revolutionary tool for improving the efficiency, productivity and sustainability of farming practices. Technologies are shaping the way farmers manage their crops and farms, offering methods that go beyond traditional practices. AI enables automated crop monitoring through technologies such as drones and sensors. These devices collect precise data on field conditions, including moisture levels, plant health and the presence of pests. For example, using multi-spectral cameras, AI-powered systems can detect water stress in plants, enabling farmers to react quickly. [345]Thanks to AI, agricultural yields can be predicted with greater accuracy. By combining historical data on weather, soils and farming practices, machine learning models can estimate potential yields. This helps farmers make informed decisions on crop planning and resource optimization. AI plays a crucial role in managing inputs such as water, fertilizers and pesticides. By analyzing real-time data, farmers can optimize the use of these resources. For example, smart irrigation systems can adjust the amount of water applied according to specific crop needs, resulting in substantial water savings. AI-powered machines, such as autonomous tractors and harvesting robots, are revolutionizing the way agricultural work is done. These technologies not only reduce the labor required, but also improve the precision and efficiency of agricultural operations, such as planting, cultivation and harvesting. [346]AI facilitates the analysis of large quantities of

[344] www.ultralytics.com
[345] www.intelligence-artificielle-school.com
[346] Artificial intelligence and agriculture;

agronomic data from a variety of sources (sensors, satellites, etc.). Using advanced algorithms, it is possible to identify patterns and correlations that can inform management decisions. This helps farmers understand the complex interactions between crops and their environment. AI is used to develop systems for detecting and preventing plant diseases. By analyzing images of crops, algorithms can identify early signs of disease or pest infestation. This enables proactive treatment, reducing the use of chemicals and minimizing crop losses.

Numerous applications and platforms use AI to offer integrated farm management solutions. These tools enable farmers to monitor crop performance, plan operations and access market information, while facilitating communication with other stakeholders. [347]Artificial intelligence is transforming agriculture by offering innovative solutions to improve productivity and sustainability. Its many applications, from crop monitoring to task automation, enable farmers to adapt to today's industry challenges, including increasing pressure on resources and the need to cater for a growing global population. By embracing these technologies, the future of agriculture looks bright, with methods that not only increase yields, but will also be more respectful of the environment.

[347] Idem.

3-1-2- Concrete examples of AI use in precision agriculture

Here are a few concrete examples of the use of artificial intelligence in precision agriculture.

1. Drones and aerial imaging

Example: Startup **Precision Hawk** uses sensor-equipped drones to capture multi-spectral images of fields. Using AI, the images are analyzed to detect disease and water stress in crops. The algorithm quickly identifies problem areas, enabling farmers to target interventions.[348]

2. Intelligent Irrigation Systems

Example: CropX, an Israeli company, offers an intelligent irrigation solution that uses sensors to monitor soil moisture. AI software analyzes this data in real time and provides recommendations on how much water to apply, optimizing water use while maintaining crop health .[349]

3. Agricultural robots

Example: Harvest Croo Robotics has developed a harvesting robot that uses AI technologies to identify and harvest strawberries. The robot is able to distinguish between ripe and unripe fruit, allowing only fruit that is ready to be harvested to be picked, thus increasing efficiency and reducing waste.[350]

---

[348] PrecisionHawk.(2020). "How PrecisionHawk Uses AI in Agriculture."

[349] Id.

[350] Harvest Croo Robotics (2021). "Strawberry Harvesting Robot."

- Disease Detection

Example: Plantix is a mobile application that uses AI to diagnose plant diseases based on photos taken by farmers. Deep learning algorithms analyze images, identify diseases and suggest treatment solutions, enabling rapid intervention .[351]

Figure15 : Crop disease detection

- Yield prediction

Example: The Climate Corporation, a Bayer subsidiary, has developed a predictive model that uses weather data, soil analysis and yield history to provide crop yield forecasts. This model helps farmers plan their operations and make decisions about which crops to plant.[352]

-Soil Fertility Management

Example: AgBiome uses AI to develop biological plant disease control products by analyzing soil microorganisms. Their technology identifies beneficial microbiomes that promote soil and crop health,

[351] Plantix.(2020). "The Plant Doctor in Your Pocket."

[352] The Climate Corporation. (2021). "Unlocking the Power of Data for Farmers."

helping to reduce reliance on chemicals[353] . These examples illustrate how AI is being integrated into precision agriculture to improve farm efficiency, productivity and sustainability. By combining modern technologies with traditional farming practices, it is possible to meet growing food demand while preserving the environment.

3-2- Impact of AI on the efficiency of precision agriculture

The introduction of artificial intelligence (AI) in this field has brought about a radical transformation. AI makes it possible to analyze large quantities of data, make informed decisions and implement actions in real time. Precision agriculture relies on the use of advanced management techniques that integrate technologies such as geolocation (GPS), sensors, drones and data platforms. These tools enable the collection of precise information on soil conditions, plant health, and climatic conditions. [354]Integrating AI into precision agriculture not only enables data to be collected, but also analyzed and relevant conclusions to be drawn. Machine learning and deep learning algorithms are used to identify patterns in the data, predict yields, and automate processes. The impact of AI on the efficiency of precision agriculture is manifested through several significant benefits.

One of the main benefits of AI is increased yields. Using precise analyses of harvested data, farmers can identify the optimum time to irrigate, fertilize and harvest. For example, **The Climate Corporation**

[353] AgBiome (2021). "Using Microbes to Improve Soil Health."

[354] Agriculture and AI;

is developing models that predict yields by analyzing various parameters, helping to maximize production. Increased efficiency thanks to AI also helps reduce operational costs. For example, smart irrigation systems, such as those developed by **CropX**, adjust watering based on actual crop needs, significantly reducing water and energy costs. By optimizing the use of resources, AI also contributes to the sustainability of agriculture. The reduced use of pesticides and fertilizers, thanks to targeted application generated by algorithms, minimizes environmental impact. Research has shown that the use of AI can increase input efficiency by 10-30% (Liu et al., 2018) .[355]

[356]Despite the undeniable benefits, integrating AI into precision agriculture also poses challenges. The initial cost of deploying AI-based technologies can be prohibitive for many farmers, particularly in developing countries. Investments in equipment, software and training are needed before the benefits are visible. AI requires large quantities of high-quality data to operate effectively. Farmers must therefore invest in the collection of reliable and accurate data, which can be an obstacle. Implementing AI requires a certain amount of technological mastery. Many farmers lack the appropriate training to use these technologies, which limits their access to the potential benefits.

---

[355] Liu, Y., Zhang, Y., & Smith, P. (2018). "The Use of Precision Agriculture Technologies to Guide Input Management and Reduce Environmental Impact". *Agricultural Systems*, 164, 198-205.

[356] Idem.

[357]The impact of artificial intelligence on the effectiveness of precision agriculture is profound and multidimensional. Thanks to innovative technologies, AI is not only transforming the way farmers implement their practices, but also offering opportunities to make agriculture more sustainable and beneficial to the environment. However, addressing the challenges of integrating these technologies is crucial to maximizing their benefits. With appropriate investment in technology and training, AI-powered precision farming could represent the future of sustainable and efficient agriculture.

3-2-1-Analysis of productivity gains

[358]The introduction of artificial intelligence (AI) in precision farming is having a significant impact on productivity gains. Advanced technologies such as machine learning, data analysis, and smart sensors enable farmers to optimize their farming operations in a number of ways. Here are some key aspects of this analysis:

[359]The use of AI enables in-depth analysis of agronomic data. Machine learning algorithms can process large amounts of data from various sources (such as soil sensors, satellite images, and drones) to provide real-time recommendations concerning crop management, irrigation, and fertilization. AI also makes it possible to optimize the use of agricultural resources. Thanks to techniques like predictive analytics,

---

[357] Id.

[358] How is AI being used in agriculture? en.isahit.com

[359] Wolfert, S., Ge, L., Verdouw, C., & Bogaardt, M.-J. (2017). "Big Data in Smart Farming - A review." *Agricultural Systems*, 153, 69-80.

farmers can better manage water, fertilizers, and pesticides. This not only reduces costs, but also minimizes environmental impact[360] . The integration of sensors and IoT (Internet of Things) devices enables continuous monitoring of crops and fields. This helps to detect diseases, infestations or other potential problems more quickly, enabling early interventions . [361][362]AI tools enable farmers to implement specific practices such as variable input management. This means that inputs (water, fertilizers, pesticides) are applied according to the specific needs of areas of the field, increasing efficiency and reducing waste.

A well-orchestrated application of AI can lead to a significant increase in yields. Studies have shown that farmers who adopt AI-based technologies can achieve above-average yields due to optimized management and use of accurate information[363] . The impact of AI on precision agriculture is evident through optimized decision-making, efficient resource management, constant crop monitoring, precise interventions, and increased yields. However, it is crucial for farmers to familiarize themselves with these technologies and assess the costs and benefits associated with their adoption. The future of agriculture

---

[360] Liakos, K. G., d'Auria, T., Pappas, C., & Pappas, G. (2018). "Machine Learning in Agriculture: A Review." *Sensors*, 18(8).

[361] Id.

[362] Pinder, M. A., et al. (2020). "Smart farming and technologies: A review of the latest developments." *Computers and Electronics in Agriculture*, 173, 105431.

[363] Id.

lies firmly in the integration of AI into its practices, with promises of sustainability and increased efficiency .[364]

3-2-2-Reducing operating costs and optimizing resources

[365]Precision farming, incorporating advanced technologies, aims to increase the efficiency and sustainability of agricultural systems. Artificial intelligence (AI) plays a key role in this transformation, helping to reduce operating costs and optimize resources. As the world's population grows, demand for food is increasing, while resources such as water and arable land are becoming increasingly limited. Precision agriculture must therefore meet this growing demand while using resources more efficiently. AI offers promising solutions to these challenges, making farms less vulnerable economically and ecologically. [366]Artificial intelligence can reduce operating costs by optimizing input management. Thanks to sophisticated algorithms, farmers can accurately determine the amount of water, fertilizer and pesticides required for each plot of land, based on the specific characteristics of the crop and environmental conditions. A study by Zhang and Wang (2019) showed that using AI for input management in corn crops enabled farmers to reduce their fertilizer costs by 20% while maintaining high yield levels. AI-powered autonomous robotics and intelligent farm vehicles are also helping to reduce labor costs. Autonomous machines capable

[364] European Commission (2020).

[365] Id.

[366] How is artificial intelligence transforming the electricity sector? cgspace.cgiar.org

of sowing, irrigating or harvesting reduce dependence on human labor, particularly in regions where labor costs are high.

As an example, agricultural technology company *Blue River Technology* has developed a precision spraying system that uses AI to identify weeds and apply herbicides only to them, significantly reducing chemical costs while improving operational efficiency (Blue River Technology, 2021)[367] . One of the major challenges facing agriculture is water management. By using sensors to monitor soil moisture and weather data in real time, AI can help farmers irrigate more efficiently, saving water and reducing irrigation costs. A pilot project in California demonstrated that using AI-based smart irrigation systems could save up to 30% water and reduce water costs by around 25% (California Department of Water Resources, 2020) .[368]

AI also helps optimize fertilizer use. Soil analysis and moisture data enable nutrients to be applied in a precise and targeted manner, avoiding the waste and costs associated with over-application. According to a report by Wolfert et al. (2017)[369] , precision farming technologies incorporating AI have enabled farmers to reduce fertilizer use by 15-25%, resulting in a substantial drop in production costs. Integrating AI with farm management platforms enables efficient coordination between different farming operations.

[367] Blue River Technology (2021). "Our Technology: See How We're Redefining Agriculture."

[368] California Department of Water Resources. (2020). "California's Water Use Approaches."

[369] Wolfert, S., Ge, L., Verdouw, C., & Bogaardt, M.-J. (2017). "Big Data in Smart Farming - A review." *Agricultural Systems*, 153, 69-80.

Intelligent software can analyze harvested data and provide recommendations to farmers on the best practices to adopt. Data analysts play a crucial role in this optimization. AI can identify trends and anomalies in crop performance, going so far as to predict future yields and adjust accordingly to maximize profits. The impact of AI on precision agriculture is profound, offering effective solutions for reducing operating costs and optimizing resource use.

3-3-Case study on the impact of AI on precision agriculture in Africa

[370]Africa faces many challenges in agriculture, including variable climatic conditions, inadequate resource management, rapid population growth, and persistent food insecurity. There are many practical cases of the use of AI in agriculture, and results are being drawn.

3-3-1-Practical use of AI in agricultural projects in Africa

[371]*M-KOPA* is a company that uses AI and mobile technology to provide solar energy systems to Kenyan farmers. This service gives farmers access to automatic irrigation devices, reducing their dependence on water while increasing agricultural production. Through the use of climate data and soil sensors, *M-KOPA* has reduced irrigation costs by 30% and increased crop yields by 26% (M-KOPA, 2021).

---

[370] Id.

[371] M-KOPA. (2021). "Transforming Agriculture with Solar Power in Kenya."

From South Africa, *AgriPredict* is a platform that uses AI to create predictive models about crop yields and potential diseases. By analyzing weather data, satellite images and other parameters, *AgriPredict* helps farmers make informed decisions. According to a study conducted by *AgriPredict*, farmers using the platform recorded a 20% increase in yields and a reduction in disease losses of up to 40% (AgriPredict, 2020) .[372]

From Nigeria, *FarmCrowdy* is a participatory funding platform for agriculture that uses AI to help investors choose the most promising agricultural projects. The platform collects data on yields, market conditions and agricultural trends. By facilitating access to financing and optimizing crop choices, *FarmCrowdy* has helped increase partner farmers' profits by an average of 30% (FarmCrowdy, 2019)[373] . The use of drones to monitor crops is booming in Africa. Drones equipped with multi-spectral cameras can capture images of fields, enabling accurate assessment of crop health.

- **Example**: In Ghana, projects have been set up to use drones to manage cocoa crops, improving early detection of disease and leading to higher crop yields.

AI-based management systems analyze data from a variety of sources, such as soil sensors and weather forecasts, to provide recommendations on irrigation and fertilization.

---

[372] AgriPredict (2020). "Leveraging AI for Enhanced Agricultural Productivity."

[373] FarmCrowdy (2019). "Investing in Food Security: A Report on Our Impact."

- **Example**: *Farming Data* in Kenya collects and analyzes data to advise farmers on the best use of water and fertilizers, increasing efficiency and yields.

[374]The integration of artificial intelligence into precision agriculture in Africa is testimony to the potential of this technology to transform agricultural production methods. [375]Initiatives such as *M-KOPA*, *AgriPredict* and *FarmCrowdy* illustrate how AI can improve yields, reduce operating costs and contribute to food security. However, to fully realize these benefits, it is essential to raise awareness, improve access to technologies and foster effective collaboration between the various players in the sector.

3-3-2-Results and lessons learned

[376]The application of artificial intelligence (AI) in precision agriculture in Africa has led to significant results, both economically and in terms of environmental sustainability. Numerous projects using AI have shown improved crop yields. For example:

- **AgriPredict** in South Africa has helped farmers increase yields by 20% thanks to predictions based on climate and crop data. The use of AI techniques to anticipate disease and weather conditions has enabled more proactive crop management.

---

[374] Idem.
[375] Id.
[376] Id.

- In Kenya, smart irrigation systems, supported by real-time data analysis, have led to 30% higher yields in maize and vegetable plantations (M-KOPA, 2021).

[377]The application of AI in input management, including water, fertilizers and pesticides, has reduced production costs. Companies such as *Blue River Technology* have demonstrated a 20-30% reduction in chemical costs through the targeted application of herbicides only to weed-infested areas. AI has also facilitated crop diversification, helping to strengthen the resilience of agricultural systems in the face of climate change. [378]The results obtained following the application of artificial intelligence in precision agriculture in Africa testify to significant benefits in terms of increased yields, reduced costs, better resource management and lower post-harvest losses.

These findings illustrate not only the potential of AI to transform agriculture on the continent, but also the need to invest in infrastructure and farmer training to maximize these benefits. As the adoption of AI technologies becomes more widespread, Africa could experience an agricultural revolution that supports food security and environmental sustainability.

---

[377] www.fao.org

[378] Farming Data Initiative. (2022). "Reducing Post-Harvest Losses with Data-Driven Decisions." [Online].

Chapter 4:

Challenges and prospects for AI integration in agriculture

[379]Agriculture, a fundamental pillar of the global economy and food security, is facing unprecedented challenges, such as climate change, population growth and dwindling natural resources. Against this backdrop, artificial intelligence (AI) is emerging as a promising solution capable of revolutionizing agricultural practices. AI systems can optimize resource use, improve crop yields, and monitor soil health while minimizing environmental impact. [380]With technologies such as machine learning, computer vision and predictive analytics, AI enables real-time data collection and informed decision-making. For example, farmers are using sensor-equipped drones to monitor crop health, enabling targeted intervention rather than a one-size-fits-all treatment approach (Zhang & Kovacs, 2012). Similarly, AI systems can analyze weather conditions and historical data to predict crop yields and adjust farming practices accordingly (Liakos et al., 2018).[381]

However, the adoption of AI in agriculture presents several challenges. These include technology integration issues, farmer skills, data accessibility, and ethical considerations related to the privatization of agricultural data. In addition, the impacts of AI on

[379] Journalopenedition.org.
[380] Id.
[381] Liakos, K. G., Busato, P., Moshou, D., & Tzamalouka, G. (2018). "Machine Learning in Agriculture: A Review." *Sensors*, 18(8), 2674.

smallholders and food security in vulnerable regions need to be thoroughly examined .[382]

In short, while AI offers promising prospects for transforming agriculture, a thoughtful and inclusive approach will be essential to overcome the challenges ahead.

4-1-Challenges of AI integration

Artificial intelligence (AI) is revolutionizing many sectors, including agriculture. Innovative techniques such as machine learning, robotics and data analysis are helping to optimize yields, manage resources more efficiently and anticipate crop disease. Despite the promising benefits, the integration of AI into the agricultural sector faces significant challenges. One of the major challenges to the integration of AI in agriculture lies in the uneven technological infrastructure. In many rural areas, particularly in developing countries, access to broadband Internet and advanced technological devices is limited. [383]A study by McKinsey & Company (2019) highlights that the lack of digital connectivity hinders the development of agricultural technology and limits the use of AI-based tools. [384]Farmers often rely on aging infrastructure, which hampers their ability to adopt AI solutions. Consequently, substantial investment in infrastructure is needed to ensure that farmers have access to the tools

[382] Agyekum, K., Frimpong, S., & Gyasi, E. A. (2021). "Artificial Intelligence in Agriculture: Challenges and Prospects." *Journal of Agricultural Science and Technology*, 23(2), 123-137.

[383] Id.

[384] Artificial intelligence in agriculture;

and technologies they need to reap the benefits of AI. This raises the issue of equity in access to technology; smallholders, who account for a significant share of global agricultural production, risk being left behind without adequate support. [385]To meet this challenge, it is crucial to set up targeted training programs that enable farmers to familiarize themselves with AI tools. Community-wide awareness-raising initiatives can also play a vital role in helping farmers understand the benefits of AI and develop the skills needed to use it effectively. In addition, collaborations between government organizations, technology companies and academic institutions can also promote education and awareness. Data collection and use is another major obstacle to the integration of AI in agriculture. AI systems rely on large and diverse datasets to generate useful analyses and recommendations. However, access to quality data, particularly in contexts where agricultural practices are informal, can be difficult . [386]

Data confidentiality and security issues also arise. Farmers may be reluctant to share their data, fearing misuse or loss of intellectual property. The development of clear regulations and ethical practices concerning data collection and use is essential to guarantee a climate of trust. This requires collaboration between government actors, companies and farmers (Wang et al., 2020)[387] . The integration of AI into agriculture also raises ethical issues. Increased use of technology can lead to over-reliance on automated solutions, reducing farmers'

[385] Ibid.

[386] Id.

[387] Wang et al ; (2020).

resilience to hazards. Furthermore, there are fears that AI will accentuate existing inequalities in wealth and resources, with large companies finding it easier to adopt these technologies compared to smallholders (Agyekum et al., 2021) .[388]

It is therefore vital that AI integration initiatives are designed with an ethical approach. This includes engaging farmers in the technology development process, ensuring that solutions truly meet their needs and contexts. A well-defined "AI ethic" must be adopted to ensure that technological innovation benefits everyone and does not create new imbalances.

Finally, the impact of AI on food security and the sustainability of agriculture is a crucial aspect to consider. The deployment of AI technologies could exacerbate the over-consumption of natural resources if appropriate regulations are not put in place. The use of AI to optimize yields must be balanced by sustainable agricultural practices that preserve the environment. Agricultural sustainability is not just about productivity, but also about healthy ecosystems and resilient food systems. The implementation of AI must therefore be accompanied by reflection on its long-term impacts and efforts to promote responsible, environmentally-friendly agriculture (Schmidt et al., 2020)[389] . Integrating artificial intelligence into agriculture presents colossal challenges that cannot be ignored. The lack of

---

[388] Id.

[389] Schmidt, J., Winter, S., & Jensen, M. (2020). "Artificial Intelligence in Agriculture: Addressing the Threats to Food Security and Sustainability." *Agriculture and Human Values*, 37(2), 477-491.

infrastructure, the need for adequate skills, concerns raised by data, ethical considerations, and the potential impact on food security are all issues that require sustained attention. Nevertheless, with a collaborative and thoughtful approach, it is possible to overcome these challenges and leverage the benefits of AI to build a more sustainable and resilient future for agriculture.

4-1-1-Digital infrastructure

[390]Modern agriculture is increasingly interconnected and reliant on digital technologies to improve efficiency, productivity and sustainability. Technologies such as precision farming, geographic information systems (GIS), and the Internet of Things (IoT) are revolutionizing agricultural practices. However, the lack of an adequate digital infrastructure is one of the main challenges to the adoption and integration of these technologies in the agricultural sector. One of the main digital infrastructure issues in agriculture is unequal access to digital technologies. In many rural areas, particularly in developing countries, access to the Internet remains limited. According to a report by the International Telecommunication Union (ITU, 2020), some 3.7 billion people worldwide have no access to the Internet, most of them in rural areas. This situation poses a significant barrier to the adoption of digital agricultural practices that could improve yields and sustainability. [391]Initiatives to improve

[390] Id.

[391] www.fao.org

Internet access in rural areas, such as the expansion of telecoms networks and subsidy programs for digital infrastructure, are needed.

However, these efforts can require substantial investment and coordination between various stakeholders, including governments, telecommunications companies and non-governmental organizations (NGOs).

What's more, Internet connectivity in rural areas is not just a question of access, but also of quality. Even in areas where the Internet is available, connection speed and reliability may be insufficient to support advanced digital applications. Studies show that accurate and reliable Internet speeds are essential for the efficient operation of precision farming technologies (Meyer et al., 2019)[392] . The lack of this adequate connectivity is particularly problematic during critical periods, such as sowing or harvesting, when real-time data is crucial. The lack of reliable connectivity also limits farmers' ability to use cloud-based applications, which can provide real-time data analysis and recommendations. Alternative solutions, such as mesh networks and low-energy communication technologies, need to be explored to improve connectivity in remote areas (Bertot et al., 2019). [393] [394]Digital infrastructure requires significant initial investment, and

[392] Meyer, E., Fuchs, C., & Hohfeld, L. (2019). "Digital Transformation in Agriculture: Current Adoption Trends and Future Perspectives." *Journal of Agricultural Engineering*, 50(3), 132-142.

[393]Bertot, J. C., Jaeger, P. T., & Grimes, J. M. (2019). "The Role of The Internet and Communication Technologies in the Transformation of Agriculture." *International Journal of Information Systems for Crisis Response and Management*, 11(1), 30-50.

[394] Idem.

unequal access to technologies can exacerbate existing economic disparities. Large farms, which have access to the necessary financial resources and skills, can easily adopt digital technologies. In contrast, small farmers, often limited by tight budgets, may be unable to make this investment, raising concerns about equity and sustainability. Studies have shown that small farms could benefit from the adoption of digital technologies, but that financial barriers to entry exist (Poudel et al., 2020)[395]. To overcome these barriers, governments and financial institutions should consider innovative financing models, such as grants, low-interest loans, and public-private partnerships to help farmers invest in digital infrastructure. Rapid technological development brings additional challenges. Technologies evolve at a rapid pace, and it can be difficult for small farms to keep up with these changes. Farmers can invest in digital solutions, but these can quickly become obsolete if infrastructures are not constantly updated. This raises questions about the sustainability of investments made in these infrastructures. [396]It is therefore crucial to foster an integrated approach that ensures that digital infrastructures remain up-to-date and adapted to farmers' needs. This could include feedback mechanisms through which farmers can express their needs and concerns, thus contributing to the evolution of technologies to meet field requirements. Digital infrastructure is essential for the

---

[395] Poudel, B., Lantagne, S., & Sabourin, E. (2020). "Barriers to Adoption of Digital Technologies in Agriculture: A Case Study of Smallholder Farmers in Nepal." *Technology in Society*, 62, 101297.

[396] Id.

transformation of modern agriculture; nevertheless, it presents many challenges that need to be overcome. Accessibility, connectivity, lack of digital skills, economic implications and sustainability issues all require attention. By adopting a collaborative approach between governments, businesses, NGOs and farmers, it is possible to invest in digital infrastructures that promote inclusivity and sustainability. The future of agriculture will depend, in large part, on our ability to solve these challenges and create a fertile digital environment for farmers worldwide.

4-1-2-Cost and accessibility of technologies

Modern agriculture increasingly relies on the use of advanced technologies to maximize yields, improve sustainability and meet the challenges of global food security. However, the integration of these technologies into farming practices is highly dependent on their cost and accessibility, factors which can create significant disparities between farmers, particularly between large-scale and small-scale operations. [397]The cost of agricultural technologies varies considerably according to the type of solution envisaged. Investments in sophisticated equipment such as drones, sensors, smart irrigation systems and management software can represent significant expenses for farmers. For example, a study conducted by the Boston Consulting Group (2019) indicates that the adoption of digital technologies in

[397] Boston Consulting Group (2019). "The Future of Agriculture: How Digital Technologies are Transforming Agriculture."

agriculture can require initial investments ranging from a few thousand to several million dollars, depending on the size of the farm and the sophistication of the technologies. The direct costs of acquiring technological equipment are only part of the total. Farmers must also consider the costs of maintenance, updating and training to use these technologies effectively. [398]According to an FAO report (2021), many farmers are reluctant to invest in digital technologies for fear of not getting a sufficient return on investment, especially if maintenance costs are high or the technologies become obsolete quickly. [399]For family farms and small producers, these costs can represent a proportionally larger share of their budget. In addition, many of them have limited access to credit, making it even more difficult to acquire advanced technologies (Kumar et al., 2020). Smallholders may also face higher financial risks due to their reliance on traditional farming practices, making the adoption of new technologies even more difficult.

The accessibility of agricultural technologies is influenced by several factors, including local infrastructure, the availability of support services, and the skills of potential users. One of the main obstacles to technology accessibility is digital infrastructure. Many rural areas lack adequate Internet access, limiting farmers' ability to take advantage of cloud-based digital solutions. According to a report by the International Telecommunication Union (ITU, 2020), around

[398] FAO (2021).

[399] Op.cit.

3.7 billion people do not have access to the Internet, mainly in rural areas. This situation makes it difficult to adopt modern cultivation techniques that depend on real-time data. Another essential aspect of accessibility is technical support. The adoption of new technologies often requires appropriate training to ensure that farmers can use them effectively. However, many farmers do not benefit from adequate training programs to familiarize themselves with these technologies. A study conducted by FAO (2021) indicates that inequalities in technological training are often exacerbated by socio-economic factors, further limiting access.

[400]Economic disparities between farmers can also lead to inequalities in access to technologies. Large farms generally have the financial resources to invest in advanced technologies and benefit from economies of scale, while smallholders may find themselves locked into traditional practices, for fear of taking on additional financial risks. This situation fosters a technology gap that can impact on the productivity and competitiveness of local farms (Poudel et al., 2020).

To improve access to agricultural technologies, several strategies can be implemented:

It is essential to explore innovative financing models to help smallholders acquire technologies. This could include subsidies, production cooperatives that purchase equipment collectively, or

[400] Poudel, B., Lantagne, S., & Sabourin, E. (2020). "Barriers to Adoption of Digital Technologies in Agriculture: A Case Study of Smallholder Farmers in Nepal." *Technology in Society*, 62, 101297.

public-private partnerships that facilitate access to financial resources[401] . Access to low-interest credit could also help farmers invest in technologies without creating undue financial pressure. Investing in digital infrastructure is crucial to ensuring the accessibility of agricultural technologies. Governments can play a major role in developing telecommunications networks in rural areas, facilitating access to cloud-based applications and other digital tools. In addition to connectivity, efforts must focus on creating an appropriate hardware infrastructure to support the use of advanced technologies.

[402]The cost and accessibility of agricultural technologies represent major challenges for the transformation of modern agriculture. While digital technologies offer significant opportunities to improve productivity and sustainability, disparities in cost and access pose obstacles to their widespread adoption. By implementing appropriate strategies, investing in digital infrastructure and empowering farmers, these challenges can be overcome to foster inclusive and sustainable agriculture.

4-1-3- Lack of technical expertise

[403]Artificial intelligence (AI) has the potential to transform a variety of sectors, including healthcare, finance, education and

---

401 Idem.

402 ITU. (2020). "Measuring Digital Development: Facts and Figures 2020." International Telecommunication Union (ITU) report.

403 Id.

agriculture. Unfortunately, the integration of these advanced technologies is often hampered by a lack of technical competence. The complexity of AI systems, the speed of technological innovation and the training requirements make it difficult for many professionals to adopt these innovations. Technical skills are defined as the set of specific knowledge and abilities required to use advanced technologies effectively. In the context of AI, this includes skills in programming, statistics, data analysis, and understanding machine learning algorithms. Technical competence is therefore a crucial element in taking full advantage of the opportunities offered by AI (Brynjolfsson & McAfee, 2014)[404] . One of the main factors contributing to the lack of technical skills is insufficient training in higher education institutions. Academic programs often fail to keep pace with technological advances. Many universities and technical schools do not yet integrate AI concepts into their curricula, leaving students ill-prepared to meet the challenges of the professional world[405] . AI is a constantly evolving field, making it difficult for professionals to stay up to date. New techniques emerge regularly, and the AI technology ecosystem changes rapidly. As a result, even professionals who have acquired technical skills may find that these

[404] Brynjolfsson, E., & McAfee, A. (2014). *The Second Machine Age: Work, Progress, and Prosperity in a Time of Brilliant Technologies*. W. W. Norton & Company.

[405] Chui, M., Manyika, J., & Miremadi, M. (2016). "Where machines could replace humans-and where they can't (yet)." *McKinsey Quarterly*.

become obsolete due to rapid advances in technology[406] . Limited access to quality training resources is also a major obstacle. In many regions, particularly in developing countries, financial interests or inadequate infrastructure can make it difficult to access online courses, workshops, or specialized training (World Economic Forum, 2020)[407] . Lack of awareness of AI's benefits and potential within organizations can also prevent effective adoption. Decision-makers and managers, particularly in small and medium-sized enterprises (SMEs), may not realize the importance of investing in AI skills development, limiting the chances of successful integration[408] . Lack of technical skills leads to delayed adoption of AI technologies. Organizations that cannot train their staff are often left behind in an increasingly digital economy. This has consequences not only for the companies themselves, who lose competitive opportunities, but also for the economy as a whole[409] . Inefficiency resulting from lack of technical competence can lead to lower productivity. Workers are unable to make optimal use of available AI tools, resulting in wasted resources and suboptimal performance (Kaplan & Haenlein, 2019)[410] . When organizations can't fully exploit the potential of AI, it compromises

---

[406] T. H., & Ronanki, R. (2018). "AI for Davenport the Real World." *Harvard Business Review*, 96(1), 108-116.

[407] World Economics forum 2020.

[408] Id.

[409] Id.

[410] Kaplan, J., & Haenlein, M. (2019). "Siri, Siri, in my Hand: Who's the Fairest in the Land? On the Interplay of Trust and AI in User-Technology Interaction." *Business Horizons*, 62(3), 375-384.

their growth and ability to innovate. The technology gap between well-established companies and those lacking AI skills could also exacerbate inequalities. Large companies often have the resources to train their staff or hire experts, while small businesses are left in the dark, which can have a negative impact on job creation and economic dynamism.

To combat the lack of technical skills, it's crucial to strengthen educational programs. Higher education institutions need to update their curricula to include courses on AI, machine learning, and data science. Setting up partnerships between universities and companies can also foster innovation in training programs, ensuring that they meet market needs. Companies also need to invest in ongoing training and professional development for their employees. This can include workshops, online courses, and conferences on AI-related topics. By integrating education throughout a career, workers can acquire the skills they need in the face of rapid technological change (Davenport & Ronanki, 2018)[411] . Encouraging a culture of innovation within organizations is essential. The emphasis should be on experimentation and the acceptance of failure as a learning step. This can help create an environment where employees feel comfortable exploring and adopting new technologies (Müller & Bostrom, 2016)[412] . Promoting the use of e-learning platforms is a practical solution to overcoming barriers to accessing training. Numerous resources are now available

---

[411] Id.

[412] Müller, V. C., & Bostrom, N. (2016). "Future Progress in Artificial Intelligence: A Survey of Expert Opinion." *Global Priorities*

for learning AI at one's own pace, making training more accessible for those unable to attend face-to-face courses. The lack of technical skills is a major challenge to the integration of artificial intelligence in various sectors. This challenge is fuelled by inadequate academic training, limited access to training resources, and rapidly evolving technologies. The consequences are significant, ranging from delayed adoption to increased risk of economic inequality.

On the other hand, solutions do exist, which require collaboration between educational institutions, businesses and governments to help overcome this challenge. Investing in the development of AI skills is not only beneficial for individuals and businesses, but also essential to ensuring a sustainable future in an increasingly digitized economy.

4-2-Perspective and recommendations

[413]The integration of artificial intelligence (AI) in various sectors, including agriculture, offers promising prospects for improving business processes, boosting productivity and promoting sustainability. As agriculture faces growing challenges such as population growth, climate change and changing consumer demands, AI is emerging as an innovative solution capable of optimizing resources and transforming traditional farming practices. AI systems can analyze vast data sets, predict crop yields, manage water resources and even improve plant resistance to disease. However, to take full

---

[413] Putting artificial intelligence to work in agriculture;

advantage of these opportunities, it is essential to identify and overcome the obstacles to the integration of these technologies, through good policies.

4-2-1- Public policies to encourage AI adoption

Artificial intelligence (AI) is redefining the contours of many sectors, including agriculture, healthcare, transportation and education. However, its widespread adoption requires a public policy framework that can both encourage adoption and reduce potential barriers. Governments play a crucial role in implementing policies that promote the integration of AI into the economy, by offering incentives, investing in education and training, and developing suitable infrastructures. [414]To encourage the adoption of AI, it is essential to develop a clear and appropriate regulatory framework. This involves establishing laws that define safety and ethical standards for the use of AI.

For example, the European Union is working on a proposal for an AI regulation that aims to establish clear rules for users and developers, while guaranteeing safety and the protection of fundamental rights (European Commission, 2021)[415] . The effective use of AI relies on high-quality data. Governments must therefore put in place policies that protect personal data while promoting its use for innovation. For example, the General Data Protection Regulation (GDPR) in Europe sets clear limits on data use, while facilitating

---

[414] Ibid.

[415] Id.

access to anonymized datasets for research and development (CNIL, 2021)[416] . Governments can encourage the adoption of AI by offering subsidies and tax credits for companies wishing to invest in AI technologies. For example, Canada's "AI for Business" program offers funding to help companies integrate artificial intelligence into their operations (Government of Canada, 2020)[417] . These financial incentives can ease the initial investment burden, making it easier for SMEs to access these technologies. Governments can also play an active role by investing in AI research and development (R&D). Initiatives such as the Horizon Europe program, which funds AI research and innovation projects, are key to advancing the state of the art in this field.

With increasing automation, some jobs may become obsolete. Governments need to set up retraining programs to help workers acquire new AI-related skills, ensuring a smooth transition to a changing labor market. [418]Integrating artificial intelligence into our society requires solid, well-adapted public policies. By developing a regulatory framework, implementing financial incentives, improving education and training, and fostering public-private collaboration, governments can create an environment conducive to innovation and ensure that AI adoption is ethical and inclusive. Best practices

---

[416] CNIL. (2021). "Règlement Général sur la Protection des Données." Commission Nationale de l'Informatique et des Libertés.

[417] Government of Canada (2020). "Artificial Intelligence for Business: Funding Opportunities." Government of Canada.

[418] Id.

observed in countries such as Canada, the United States and the nations of the European Union show that it is possible to create a dynamic ecosystem that benefits society as a whole .[419]

4-2-2-Reinforcing local capacity

Local capacity building is crucial, to ensure that farmers and rural communities are equipped with the skills, knowledge and tools they need to take advantage of this technology. Local capacity building is essential for several reasons: Local farmers need to be involved in the AI adoption process. Improving their understanding of AI technologies and providing them with the necessary tools creates an environment conducive to innovation. This enables farmers to make informed decisions that can lead to improved farming practices (Davis & He 2020)[420] . Agriculture is profoundly influenced by local factors such as climate, soil and cultural practices. Capacity building ensures that AI solutions are tailored to local specifics, maximizing their effectiveness and adoption (World Bank, 2019)[421] . Setting up training programs for farmers and young professionals can play a key role in integrating AI into agriculture. [422]Initiatives such as MOOCs

---

[419] Krause, J., et al. (2020). "Artificial Intelligence in Education: A Global Perspective." *Education and Information Technologies*, 25(2), 479-494.

[420] Davis, K., & He, Y. (2020). "The Role of Digital Agriculture in Supporting the Agriculture Sector." *Agricultural Systems*, 180, 102790.

[421] World Bank 2019.

[422] Sustainable Development Solutions Network.(2020). "Building a Sustainable Future with AI in Agriculture." SDSN.

(Massive Open Online Courses) and hands-on workshops can teach the skills needed to use AI tools, such as drones for crop monitoring or data analysis software. Awareness campaigns should be run to inform farmers of the benefits of AI and associated technologies. For example, field demonstrations can illustrate how to use AI tools to optimize irrigation or reduce post-harvest losses, boosting acceptance of these technologies (Tey and Brindal, 2012) .[423]

Local capacity building is a key element in the successful integration of artificial intelligence into agriculture. The strategies implemented must be adapted to local realities and aim to empower farmers, encourage innovative partnerships and ensure access to the necessary resources and infrastructure. By investing in the development of skills and knowledge within rural communities, it is possible to transform agriculture through AI, while promoting the sustainability and resilience of food systems.

4-2-3-Public-private partnership for the development of agricultural techniques

Agriculture is a key sector for economic development in Africa, accounting for a significant share of the gross domestic product (GDP) of many countries and employing millions of people. However, it faces many challenges, including climate change, poverty, food insecurity and lack of access to modern technologies. Public-private

[423] Tey, Y. S., & Brindal, M. (2012). "Factors Influencing the Adoption of Precision Agriculture in Australia." *Precision Agriculture*, 13, 48-71.

partnerships (PPPs) offer an innovative and effective approach to developing sustainable farming techniques, improving productivity and supporting the transition to resilient agriculture. PPPs bring together the financial, technical and human resources needed to develop and implement agricultural initiatives. By combining public and private sector capabilities, these partnerships can increase the effectiveness of investments in agriculture (World Bank, 2019)[424] . The private sector, often more agile and innovative, plays a crucial role in introducing new agricultural technologies and practices. PPPs facilitate the adoption of solutions such as precision farming, advanced irrigation technologies and data analysis systems, crucial for improving productivity (FAO, 2020) .[425]

Partnerships can also strengthen local capacity for training and knowledge transfer. Private companies can provide technical expertise, while public institutions can offer a regulatory and logistical framework. This promotes better integration of new techniques within farming communities (Makinde et al., 2021)[426] . The "Feed the Future" program, initiated by the US government, has established partnerships with private companies to improve food security in Africa. For example, collaborations with agricultural companies have led to the

[424] World Bank 2019.

[425] Id.

[426] Makinde, A. A., et al. (2021). "Public-Private Partnerships in Agricultural Development: Lessons from the African Experience." *Journal of Agriculture and Food Research*, 4, 100155.

development of disease-resistant seed varieties, improved cultivation practices and efficient distribution systems (USAID, 2020) .[427]

CAADP, an African Union initiative, aims to strengthen cooperation between governments and the private sector to develop sustainable agricultural strategies. This partnership has mobilized investment to improve agricultural infrastructure, research and development, and strengthen agricultural value chains (African Union, 2019)[428] . One of the main obstacles to effective PPPs in Africa is the lack of a clear and consistent regulatory framework. Unevenly applied laws and policies can create uncertainty for private investors and discourage their commitment (OECD, 2021)[429] . Agricultural investments, particularly in innovative technologies, entail high financial risks. PPPs need to design models that share risk equitably between public and private partners to ensure project viability. The adoption of new agricultural techniques can meet with resistance from farmers, who are often attached to traditional practices. Awareness-raising and training are essential to help farmers understand the benefits of new technologies and techniques (Wiggins & Keats, 2019)[430] . Public-private partnerships can play a crucial role in the

---

[427] USAID. (2020). "Feed the Future: Progress and Results." USAID.

[428]African Union. (2019). "Malabo Declaration on Africa's Accelerated Agricultural Growth and Transformation for Shared Prosperity and Improved Livelihoods." African Union.

[429] OECD. (2021). "Public-Private Partnerships for Sustainable Development: Guidance for the Agricultural Sector." OECD.

[430] Wiggins, S., & Keats, S. (2019). "Leverage the Change: A New Vision for Agricultural Development in Africa." *Overseas Development Institute.*

development of agricultural techniques in Africa, bringing together the necessary resources, facilitating innovation and building local capacity. Although challenges remain, examples of successful projects demonstrate the potential of PPPs to transform the agricultural sector. Continued commitment, the strengthening of regulatory frameworks and the mobilization of strong political will are essential to maximize the positive impact of these partnerships on food security and rural development in Africa.

Conclusion

Digitization represents an unprecedented opportunity for rural economies in Africa, particularly in the agricultural sector, which is the cornerstone of economic survival in many regions. The introduction and integration of artificial intelligence (AI) in precision agriculture is transforming traditional farming practices. They not only increase productivity, but also enhance sustainability, optimizing resources against a backdrop of growing environmental challenges and climate change. Digitization, through technologies such as sensors, drones, and AI-based analytical tools, improves decision-making by enabling a better understanding of soils, crops, and climatic conditions. Farmers can benefit from personalized advice in real time thanks to mobile apps that use AI to analyze local data. These innovations make it possible to monitor crop growth, detect disease early, and optimize irrigation and fertilizer use (World Bank, 2020) . [431][432]The transformation of agricultural practices is a crucial phenomenon influencing the ability of societies to meet contemporary challenges such as food security, sustainable development and climate change. This transformation, often accompanied by the adoption of new technologies and methodologies, aims to improve agricultural yields, preserve natural resources and meet the growing needs of an ever-increasing world population.

---

[431] World Bank 2020.

[432] Id.

Over the centuries, agriculture has undergone profound changes. Traditionally, farming practices were based on manual methods and local knowledge handed down from generation to generation. However, with the Green Revolution of the 1960s, developing countries began to adopt modern techniques, such as the use of chemical fertilizers and high-yielding seed varieties. This period led to a significant increase in food production, but also gave rise to environmental challenges, in particular soil degradation and depletion of water resources (Pingali, 2012)[433] . Technology plays a central role in the transformation of contemporary agricultural practices. Tools and technologies such as precision farming, drones, IoT (Internet of Things) sensors, and artificial intelligence enable precise crop monitoring and optimized resource management. For example, using AI to analyze agronomic data helps farmers make informed decisions on irrigation, fertilization and pest control. According to a study by McKinsey (2018)[434] , the adoption of these technologies could increase farm yields by 10-20% while reducing operating costs. Agriculture, a vital sector of the global economy, faces many challenges, including population growth, climate change, and the need to ensure food security. Against this backdrop, technological innovation is emerging as a key driver for transforming farming practices. Advanced techniques such as precision farming, biotechnologies and geographic information systems (GIS) are

[433]Pingali, 2012.

[434] Id.

changing the way farmers cultivate, manage their resources and respond to market needs. Precision agriculture, based on the use of technologies such as drones, sensors and geographic information systems, enables farmers to monitor and optimize their cultivation practices with greater precision. According to Aitken et al. (2016)[435] , this approach reduces the use of inputs (such as fertilizers and pesticides) while increasing yields. For example, soil sensors help assess moisture and nutrient levels, enabling more targeted irrigation and fertilization.

Biotechnology, including genetically modified (GM) seeds, plays a crucial role in crop improvement. These innovations are designed to resist disease, pests and extreme climatic conditions. Studies show that the use of GMOs has increased yields and reduced disease-related losses (Brookes and Barfoot, 2018)[436] . However, acceptance of GMOs varies considerably from region to region, raising ethical and public health concerns. GIS has become an indispensable tool for agricultural planning and management. They enable farmers to analyze geo-spatial data to make informed decisions about crop selection, water management and cultivation practices. According to an FAO (2018) report[437] , the use of GIS contributes to more efficient management of natural resources and increased agricultural productivity. Advanced technologies can significantly

[435] Aiken et al.2016.

[436] Brookes, G., & Barfoot, P. (2018). "GM Crops: Global Socio-Economic and Environmental Impacts 1996-2016." *PG Economics Ltd.*

[437] FAO Report.2018.

increase agricultural productivity. Precision agriculture, for example, contributes not only to resource optimization, but also to higher yields thanks to precise input management (Godfray et al., 2010)[438] . Technological innovation promotes sustainable agricultural practices. Reducing chemical inputs through methods such as precision agriculture helps to reduce water and soil pollution, preserve biodiversity and improve ecosystem health (Tilman et al., 2011)[439] . This is crucial for tackling contemporary ecological challenges. The increased efficiency of farming practices thanks to technologies enables farmers to reduce their production costs. For example, the use of smart irrigation systems can reduce water use, leading to considerable savings in regions where water is a precious resource (Matonkę et al., 2020) .[440]

One of the main challenges is unequal access to technology. Small-scale farmers, often in rural areas, may not have the means to invest in sophisticated equipment. Government initiatives and development programs are needed to facilitate access to these technologies[441] . The use of advanced technologies requires specific skills. It is therefore essential to train farmers in the use of these new technologies. Training and education programs are crucial to ensure

---

[438] Godfray, H. C. J., et al. (2010). "Food Security: The Challenge of Feeding 9 Billion People." *Science*, 327(5967), 812-818.

[439] Tilman et al.2011.

[440] Matonkę, B., Ksiezak, J., & Mandel, A. (2020). "Water Management in Agriculture: Importance of Irrigation Technologies." *Water*, 12(6), 1565.

[441] Id.

that farmers can take advantage of available innovations (FAO, 2018)[442] . Innovations such as GMOs raise regulatory and ethical issues. Governments need to establish clear regulations to frame the use of these technologies while addressing public concerns (Bregund et al., 2019)[443] . Acceptance of new technologies also depends on consumer confidence in their safety and efficacy.

Technological innovation is transforming farming practices, offering unprecedented opportunities to increase productivity, promote sustainability and reduce production costs. However, to maximize these benefits, it is essential to address the challenges of access, training and regulation. By integrating new technologies into farming practices, we can not only meet growing food needs, but also ensure a more sustainable and equitable agricultural future. Sustainability has become a fundamental principle in the transformation of agricultural practices. Faced with the challenges posed by climate change and increasing pressure on natural resources, farmers are adopting sustainable methods such as crop rotation, agroforestry and organic farming. These practices aim to minimize the use of chemicals, preserve biodiversity and improve soil health. [444]The FAO (2019) points out that sustainable agriculture can contribute to food security while protecting ecosystems. In addition, diversified cropping systems make farms more resilient to environmental shocks. Sustainable

---

[442] Id.

[443] Bregund, I., & Zia, M. (2019). "Consumer Acceptance of Genetically Modified Foods: A Review." *Journals of Agricultural and Environmental Ethics*, 32(2), 201-223.

[444] FAO 2019.

agriculture is a concept that aims to meet the food needs of the world's population while preserving the environment and strengthening the socio-economic systems of rural areas. Faced with growing challenges such as soil degradation, loss of biodiversity, climate change and a growing world population, sustainable agriculture is positioned as a viable solution for ensuring food security while protecting natural resources. Sustainable agriculture is defined by the Food and Agriculture Organization (FAO) as a management system that promotes economic productivity, environmental sustainability and social equity (FAO, 2014)[445] . This involves producing food in a way that minimizes environmental impacts, maximizes resource efficiency, and improves the well-being of farmers and rural communities.

The principles of sustainable agriculture are based on three main pillars:

1. **Ecological**: Protect soil, water and biodiversity, while minimizing the use of synthetic chemicals.
2. **Economic**: Ensure the financial viability of farms and promote production systems that generate sustainable income.
3. **Social**: Promoting fair and accessible working conditions, supporting local communities, and guaranteeing food security.

Sustainable agriculture helps to protect ecosystems by limiting pollution and preserving natural resources. For example, practices such as crop rotation and agroforestry improve soil health and promote

[445] FAO 2014.

biodiversity (Tilman et al., 2011)[446] . Sustainable farming systems build resilience to the impacts of climate change. By adopting techniques such as conservation agriculture, farmers can better adapt to climatic variations, preserve soil moisture and reduce erosion (Lal, 2015)[447] . Sustainable agriculture makes it possible to produce healthy, nutritious food while meeting growing demand as the world's population increases. [448]According to the FAO (2018), sustainable practices can increase agricultural yields, contributing to food security.

Sustainable agriculture can improve the living conditions of rural communities by promoting stable incomes and creating jobs. By supporting local, participatory practices, it strengthens the social and economic structures of communities. The transition to sustainable agriculture requires a willingness to change old practices and adopt new ones. This can be hampered by a lack of training, education and awareness-raising among farmers. Support from governments and institutions is essential to promote sustainable agriculture. However, there are often agricultural policies that privilege conventional approaches, neglecting the importance of sustainability (Peters et al., 2015)[449] . Agro-ecology is an approach that combines traditional

---

[446] Id.

[447] Lal, R. (2015). "Restoration of Soil Quality to Mitigate Soil Degradation." *Sustainable Agriculture Reviews*, 15, 1-19.

[448] Id.

[449] Peters, J., M. F., C. P., & M. S. (2015). "Agricultural Policies and Sustainable Development: Current Issues." *Agricultural Economics*, 46(1), 1-11.

agricultural knowledge and scientific innovations to create sustainable production systems. It focuses on biodiversity, nutrient recycling and the interaction between different species (Altieri, 2018)[450] . Integrating trees and livestock within farms, as in silvo-pastoral systems, increases biodiversity, stores carbon and provides habitats for wildlife, while generating income for farmers (Reynolds et al., 2013)[451] . Conservation agriculture emphasizes minimum soil disturbance, permanent soil cover and crop rotation. This method helps increase crop productivity while improving soil health and reducing erosion.

[452]Sustainable agriculture is an essential way of guaranteeing long-term food security while preserving natural resources and improving living conditions for farmers. Although it faces a number of challenges, including access to resources and the need to change practices, the environmental, economic and social benefits it offers cannot be ignored. It is crucial that governments, non-governmental organizations and communities join forces to promote sustainable agricultural practices and create a future where food production is both responsible and accessible to all. [453]They can support innovation, foster research and development, and offer financial incentives for the

[450] Altieri, M.A. (2018). "Agroecology: The Science of Sustainable Agriculture." *University of California Press*.

[451] Reynolds, T., T., I., & Ling, H. (2013). "Silvopastoral Systems: Creating a Sustainable Agriculture Environment." *Tropical Conservation Science*, 6(3), 367-386.

[452] Sustainable agriculture;

[453] Ibid.

adoption of sustainable practices. Initiatives such as subsidies for green technologies or training programs for farmers can contribute to the transition to more sustainable agricultural systems. [454]According to the United Nations (2021), investment in sustainable agriculture should increase significantly to meet global food needs while preserving the environment.

Social and economic changes are also driving the transformation of agricultural practices. Rapid urbanization and the migration of young people to cities are creating pressure on the agricultural sector, which must not only improve its productivity but also be more attractive to the next generation. Agricultural entrepreneurship initiatives, such as agricultural tech start-ups, are paving the way for a new dynamic, where young people can use their technology and business skills to innovate in the sector.

[455]Transforming agricultural practices is a complex, multi-dimensional process, rooted in technological innovation, sustainability, public policy support and social change. As the world's population continues to grow and environmental challenges intensify, it becomes imperative that farmers, governments and consumers work together to create a sustainable and resilient agricultural system. The future of agriculture depends on our collective ability to evolve and adapt in the face of rapid change, while preserving resources for future generations. Digitization has also opened up new avenues for farmers

---

[454] United Nations. (2021). "Transforming Food Systems for Affordable Healthy Diets."

[455] Idem.

in Africa by facilitating market access. Digital marketing platforms enable farmers to sell their produce directly to consumers, reducing dependence on often unremunerative middlemen. By using mobile payment technologies, farmers can also receive faster and more secure payment, increasing their liquidity (FAO, 2021). [456] [457]Market access is a key factor in business success, particularly in developing economies. With the emergence of the digital economy, this concept has been transformed considerably, offering new opportunities for producers, entrepreneurs and consumers alike. Digitization has broken down certain geographical and financial barriers, facilitating access to wider markets. One of the major changes brought about by the digital economy is the emergence of marketing platforms, such as Amazon, Alibaba and Etsy, which enable producers to sell their products directly to consumers without intermediaries. These platforms offer increased visibility and facilitate transactions, making products accessible to a much broader consumer base (Kumar et al., 2020)[458] . For example, farmers who once used local markets can now reach customers in other regions, or even internationally, thanks to the digitization of their operations. Digital payment systems, such as M-Pesa in East Africa, have revolutionized the way transactions are carried out. They enable simple, fast and secure payments, which is particularly important in markets where access to traditional banking services is limited. According to the International Finance Corporation

[456] FAO 2021.

[457] Idem.

[458] Kumar et al.2020.

(IFC, 2021)[459] , the adoption of digital payments has helped to increase commercial transactions and improve access to financial services for small and medium-sized enterprises (SMEs). Access to digital platforms and payment services has enabled producers and entrepreneurs to improve their business efficiency. Thanks to data analysis, they can now monitor market trends, adjust their offers accordingly and optimize their marketing strategies (Bertschek et al., 2020)[460] . For example, artisans can now use social media to promote their products, connect with potential customers, and gather feedback that helps them refine their offers.

The digital economy also fosters greater economic inclusion. By offering access to wider markets, it enables a variety of players, including women and young people, to participate in the formal economy. [461]According to a World Bank report (2020), access to digital platforms has enabled many women entrepreneurs to develop their businesses and improve their living conditions. This inclusion not only opens up economic opportunities for individuals, but also creates a positive impact on local economies.

Despite the opportunities offered by the digital economy, barriers remain. Internet access remains limited in many rural areas, and the costs of using digital technologies can be prohibitive. A report

---

[459] International Finance Corporation (IFC). (2021). "Digital Financial Services: What We Know."

[460] Bertschek, I., Kohler, W., & Niebel, T. (2020). "Digitalization and Firm Performance: A Review of the Evidence." *European Economic Review*.

[461] World Bank Report. 2020 .

by the International Telecommunication Union (ITU, 2021)[462] indicates that nearly 3.7 billion people worldwide have no access to the Internet, a major challenge for economic inclusion. Another challenge is the lack of digital skills among many producers, which limits their ability to take advantage of digital tools. Without adequate training, economic players can be left behind in this transition to the digital economy. Training programs targeting farmers and entrepreneurs on the use of digital technologies and business platforms can be crucial to maximizing the benefits of the digital economy.

To overcome barriers to digital market access, governments and private players need to invest in digital infrastructure, particularly in rural areas. The expansion of broadband Internet networks and access to affordable technologies are necessary to enable true digital inclusion[463] . Governments must also set up training and education programs to strengthen the digital skills of producers and entrepreneurs. This can include initiatives at community level to help farmers adopt digital techniques in their daily work and effectively manage their business operations online.

[464]Finally, support for digital start-ups can encourage innovation in the sector. Initiatives such as incubators and gas pedals can provide young entrepreneurs with the resources they need to develop solutions tailored to local needs, improving market access for all. Market access

---

[462]International Telecommunication Union (ITU). (2021). "Measuring digital development: Facts and figures 2021."

[463] Id.

[464] Idem.

through the digital economy brings many positive transformations for producers and consumers alike. By promoting greater economic inclusion and improving business efficiency, it offers considerable potential for economic development. However, it is crucial to overcome the challenges associated with Internet access, digital training and infrastructure. Governments and economic players need to work together to put in place favorable policies that maximize the benefits of the digital economy, ensuring that all market players can reap the benefits. Digital solutions, powered by AI, have the potential to improve financial inclusion for farmers. Thanks to data analysis, financial institutions can assess risks more effectively and offer financial products tailored to farmers' realities. This could include microloans for the purchase of modern equipment or quality seeds, needed to increase agricultural yields (International Finance Corporation, 2019).[465]

However, despite these obvious advantages, there are challenges in integrating digitalization into rural economies. A significant proportion of the rural population lacks access to electricity and the Internet, making it difficult to adopt digital technologies. In addition, there is a significant gap in digital skills, limiting the ability of many farmers to adopt new technologies. It is crucial to set up appropriate training programs to strengthen the digital skills of rural populations (UNCTAD, 2022) .[466]

---

[465]International Finance Corporation. (2019). "Financial Inclusion in Africa."

[466] UNCTAD. (2022). "The Digital Economy in Africa."

The benefits brought by digitization and AI in precision agriculture are also part of a sustainable development framework. By enabling more efficient cultivation while preserving natural resources, these innovations contribute to achieving the United Nations' Sustainable Development Goals (SDGs), notably SDG 2 (Zero Hunger) and SDG 13 (Combating Climate Change). The implementation of sustainable agriculture systems, incorporating AI, can also play a key role in mitigating the environmental impacts associated with traditional agriculture (IAASTD, 2009)[467] . To maximize the impact of digitization, it is essential to adopt an inclusive approach involving all stakeholders, including governments, NGOs, and the private sector. Proactive policies must be developed to create an environment conducive to innovation. This includes investment in digital infrastructure, the creation of public-private partnerships, and the development of a regulatory framework that promotes the safe and equitable adoption of new technologies (African Development Bank, 2020).[468]

So the impact of digitalization on rural economies in Africa, specifically through the use of artificial intelligence in precision agriculture, is immense. [469]It is a catalyst for change that can transform not only agricultural production, but also the economic and social dynamics within rural communities. However, for it to be sustainable

[467] IAASTD. (2009). "Agriculture at a Crossroads."

[468] African Development Bank. (2020). "A Digital Strategy for Africa."

[469] Idem.

and inclusive, it is necessary to overcome existing barriers, invest in education and infrastructure, and promote policies that support these transformations. The path towards a digitized and sustainable agricultural future is promising, but requires a collective commitment to ensure that the benefits are equitably distributed among all players in the sector. This work shows that value for money and informed decision-making are essential to strengthen the resilience and competitiveness of African agricultural systems in the digital age. The future of agriculture in Africa lies in the hands of technological innovation and the collective will to turn this opportunity into an engine for sustainable development.

BIBLIOGRAPHY

1. Aitken, J., Hall, R., & Ball, B. (2016). "Precision Agriculture: A Primer." *Journal of Agricultural Science.*
2. Brookes, G., & Barfoot, P. (2018). "GM Crops: Global Socio-Economic and Environmental Impacts 1996-2016." *PG Economics Ltd.*
3. FAO. (2018). "The Future of Food and Agriculture: Trends and Challenges."
4. Godfray, H. C. J., et al. (2010). "Food Security: The Challenge of Feeding 9 Billion People." *Science*, 327(5967), 812-818.
5. Matonkę, B., Ksiezak, J., & Mandel, A. (2020). "Water Management in Agriculture: Importance of Irrigation Technologies." *Water*, 12(6), 1565.
6. Tilman, D., Balzer, C., Hill, J., & Befort, B. L. (2011). "Global Food Demand and the Sustainable Intensification of Agriculture." *Proceedings of the National Academy of Sciences*, 108(50), 20260-20264.
7. World Bank. (2020). "World Development Report 2020: Data for Better Lives".
8. Bregund, I., & Zia, M. (2019). "Consumer Acceptance of Genetically Modified Foods: A Review." *Journals of Agricultural and Environmental Ethics*, 32(2), 201-223.
9. FAO. (2014). "Sustainable Agriculture for Food Security and Nutrition."
10. Tilman, D., Balzer, C., Hill, J., & Befort, B.L. (2011). "Global Food Demand and the Sustainable Intensification of Agriculture." *Proceedings of the National Academy of Sciences*, 108(50), 20260-20264.
11. Lal, R. (2015). "Restoration of Soil Quality to Mitigate Soil Degradation." *Sustainable Agriculture Reviews*, 15, 1-19.
12. FAO. (2018). "The State of Food Security and Nutrition in the World."
13. World Bank. (2020). "World Development Report 2020: Data for Better Lives."

14.Peters, J., M. F., C. P., & M. S. (2015). "Agricultural Policies and Sustainable Development: Current Issues." *Agricultural Economics*, 46(1), 1-11.

15.Altieri, M.A. (2018). "Agroecology: The Science of Sustainable Agriculture." *University of California Press*.

16.Reynolds, T., T., I., & Ling, H. (2013). "Silvopastoral Systems: Creating a Sustainable Agriculture Environment." *Tropical Conservation Science*, 6(3), 367-386.

17.Pingali, P. (2012). "Green Revolution: Impacts, Limits, and the Path Ahead." *The Proceedings of the National Academy of Sciences*, 109(31), 12302-12308.

18.McKinsey & Company (2018). "The Future of Farming: 10 Technology Trends."

19.FAO. (2019). "Sustainable Agriculture and Food Security."

20.United Nations. (2021). "Transforming Food Systems for Affordable Healthy Diets."

21.Kumar, A., Singh, S. K., & Tripathi, V. (2020). "Digital Economy: Transition and Future." *International Journal of Business and Management Studies*.

22.International Finance Corporation (IFC). (2021). "Digital Financial Services: What We Know."

23.Bertschek, I., Kohler, W., & Niebel, T. (2020). "Digitalization and Firm Performance: A Review of the Evidence." *European Economic Review*.

24.World Bank. (2020). "World Development Report 2021: Data for Better Lives."

25.International Telecommunication Union (ITU). (2021). "Measuring digital development: Facts and figures 2021."

26.World Bank. (2020). "Digitizing Agriculture for Better Outcomes."

27.FAO. (2021). "Innovation and Digital Agriculture to Tackle COVID-19."

28. International Finance Corporation. (2019). "Financial Inclusion in Africa."
29. UNCTAD. (2022). "The Digital Economy in Africa."
30. IAASTD (2009). "Agriculture at a Crossroads."
31. African Development Bank. (2020). "A Digital Strategy for Africa."
32. Aitken, J., Hall, R., & Ball, B. (2016). "Precision Agriculture: A Primer." *Journal of Agricultural Science.*
33. Brookes, G., & Barfoot, P. (2018). "GM Crops: Global Socio-Economic and Environmental Impacts 1996-2016." *PG Economics Ltd.*
34. FAO. (2018). "The Future of Food and Agriculture: Trends and Challenges."
35. Godfray, H. C. J., et al. (2010). "Food Security: The Challenge of Feeding 9 Billion People." *Science*, 327(5967), 812-818.
36. Matonkę, B., Ksiezak, J., & Mandel, A. (2020). "Water Management in Agriculture: Importance of Irrigation Technologies." *Water*, 12(6), 1565.
37. Tilman, D., Balzer, C., Hill, J., & Befort, B. L. (2011). "Global Food Demand and the Sustainable Intensification of Agriculture." *Proceedings of the National Academy of Sciences*, 108(50), 20260-20264.
38. World Bank. (2020). "World Development Report 2020: Data for Better Lives".
39. Bregund, I., & Zia, M. (2019). "Consumer Acceptance of Genetically Modified Foods: A Review." *Journals of Agricultural and Environmental Ethics*, 32(2), 201-223.
40. FAO. (2014). "Sustainable Agriculture for Food Security and Nutrition."
41. Tilman, D., Balzer, C., Hill, J., & Befort, B.L. (2011). "Global Food Demand and the Sustainable Intensification of Agriculture." *Proceedings of the National Academy of Sciences*, 108(50), 20260-20264.
42. Lal, R. (2015). "Restoration of Soil Quality to Mitigate Soil Degradation." *Sustainable Agriculture Reviews*, 15, 1-19.

43. FAO. (2018). "The State of Food Security and Nutrition in the World."
44. World Bank. (2020). "World Development Report 2020: Data for Better Lives."
45. Peters, J., M. F., C. P., & M. S. (2015). "Agricultural Policies and Sustainable Development: Current Issues." *Agricultural Economics*, 46(1), 1-11.
46. Altieri, M.A. (2018). "Agroecology: The Science of Sustainable Agriculture." *University of California Press*.
47. Reynolds, T., T., I., & Ling, H. (2013). "Silvopastoral Systems: Creating a Sustainable Agriculture Environment." *Tropical Conservation Science*, 6(3), 367-386.
48. Kumar, A., Singh, S. K., & Tripathi, V. (2020). "Digital Economy: Transition and Future." *International Journal of Business and Management Studies*.
49. International Finance Corporation (IFC). (2021). "Digital Financial Services: What We Know."
50. Bertschek, I., Kohler, W., & Niebel, T. (2020). "Digitalization and Firm Performance: A Review of the Evidence." *European Economic Review*.
51. World Bank. (2020). "World Development Report 2021: Data for Better Lives."
52. International Telecommunication Union (ITU). (2021). "Measuring digital development: Facts and figures 2021."
53. FAO. (2014). "Sustainable Agriculture for Food Security and Nutrition."
54. Tilman, D., Balzer, C., Hill, J., & Befort, B.L. (2011). "Global Food Demand and the Sustainable Intensification of Agriculture." *Proceedings of the National Academy of Sciences*, 108(50), 20260-20264.
55. Lal, R. (2015). "Restoration of Soil Quality to Mitigate Soil Degradation." *Sustainable Agriculture Reviews*, 15, 1-19.
56. FAO. (2018). "The State of Food Security and Nutrition in the World."

57. World Bank. (2020). "World Development Report 2020: Data for Better Lives."
58. Peters, J., M. F., C. P., & M. S. (2015). "Agricultural Policies and Sustainable Development: Current Issues." *Agricultural Economics*, 46(1), 1-11.
59. Altieri, M.A. (2018). "Agroecology: The Science of Sustainable Agriculture." *University of California Press.*
60. Reynolds, T., T., I., & Ling, H. (2013). "Silvopastoral Systems: Creating a Sustainable Agriculture Environment." *Tropical Conservation Science*, 6(3), 367-386.
61. Aitken, J., Hall, R., & Ball, B. (2016). "Precision Agriculture: A Primer." *Journal of Agricultural Science.*
62. Brookes, G., & Barfoot, P. (2018). "GM Crops: Global Socio-Economic and Environmental Impacts 1996-2016." *PG Economics Ltd.*
63. FAO. (2018). "The Future of Food and Agriculture: Trends and Challenges."
64. Godfray, H. C. J., et al. (2010). "Food Security: The Challenge of Feeding 9 Billion People." *Science*, 327(5967), 812-818.
65. Matonkę, B., Ksiezak, J., & Mandel, A. (2020). "Water Management in Agriculture: Importance of Irrigation Technologies." *Water*, 12(6), 1565.
66. Tilman, D., Balzer, C., Hill, J., & Befort, B. L. (2011). "Global Food Demand and the Sustainable Intensification of Agriculture." *Proceedings of the National Academy of Sciences*, 108(50), 20260-20264.
67. World Bank. (2020). "World Development Report 2020: Data for Better Lives".
68. Bregund, I., & Zia, M. (2019). "Consumer Acceptance of Genetically Modified Foods: A Review." *Journals of Agricultural and Environmental Ethics*, 32(2), 201-223.
69. African Development Bank (2020). Digitalization in Africa: an impetus for development.

70.Economic Commission for Africa (2020). Digitization in Africa.

71.McKinsey & Company (2020). The African economy: Opportunities for growth and development in a changing world.

72.World Bank (2020). Africa's Pulse: The Impact of COVID-19 on Africa's Economy.

73.FAO (2020). State of Food Security and Nutrition in the World.

74.IFAD (2020). Rural Poverty Report.

75.IPCC (2013). Climate Change 2013: The Physical Science Basis.

76.IFC (2019). Africa's Agriculture and Food Security.

77.UNDESA (2019). World Population Prospects 2019.

78.World Bank (2019). Africa's Pulse: An Update on the State of the African Economy.

79.FAO (2019). Precision Agriculture for Sustainable Development.

80.World Bank (2019). Agriculture and Food Security in Africa.

81.International Fund for Agricultural Development (IFAD) (2019). Rural Poverty Report.

82.Journal of Agricultural Science and Technology (2019). Precision Agriculture: A Review of Its Benefits and Challenges.

83.Nature Sustainability (2020). Precision agriculture for sustainable development.

84.Kamilaris, A., & Prenafeta-Boldú, F. X. (2018). *Deep Learning in Agriculture: A Survey*. Computers and Electronics in Agriculture.

85.Liakos, K. G., et al. (2018). *Agricultural robotics and the use of artificial intelligence in agriculture: A review*. Computers and Electronics in Agriculture.

86.Nassour, A., & Barnett, J. (2020). *Sustainable agriculture: Integrating precision agriculture and smart technologies*. Sustainable Agriculture Reviews.

87.Zhang, C., et al. (2019). *Digital agriculture: Transforming agriculture through digital technologies*. Journal of Agricultural and Food Information.

88.General Data Protection Regulation (GDPR): this European Union regulation, which came into force in May 2018, illustrates the importance of digitalization with regard to the protection of personal data. It establishes clear standards on digital data processing and how organizations must manage information (Regulation (EU) 2016/679).

89.Directive 2014/24/EU on public procurement: this European directive mentions digitization in the context of using electronic tools to improve public procurement procedures, indicating that digitization has become a requirement for administrative processes in the EU.

90.Loi Informatique et Libertés de 1978: In France, this law has been amended to incorporate the RGPD and provides a legal framework for the management of digital data, reinforcing the notion of digitization in a legal context.

91.Manuel Castells: In his book "The Network Society" (1996), Castells proposes the idea that digitalization modifies social structures by fostering network configurations that redefine human interactions and systems of power. In his view, the information age is fundamentally transforming social relations through digital technologies.

92.Nicolas Luhmann: In his theory of social systems, Luhmann discusses how social systems adapt and evolve in response to new communication technologies. He emphasizes that digitalization is not limited to technological tools, but also encompasses changes in social perceptions and practices.

93.Sherry Turkle: In "Alone Together" (2011), Turkle examines how digital technologies influence human interactions and social relationships. She argues that digitalization is transforming the way we communicate and interact, often to the detriment of deeper social bonds.

94. Choudary, S. P. (2017). "Platform Scale: How an Emerging Business Model Helps Startups Grow." Platform Thinking.
95. GeSI (2021). "Digital with Purpose: Delivering a SMARTer2030."
96. McKinsey (2019). "The future of work: The case for an inclusive economy."
97. McKinsey (2021). "The future of work: Reskilling and recovery."
98. Meyer, C. (2020). "How Amazon's supply chain innovation gave it an edge during the pandemic."
99. ILO (2020). "World Employment and Social Outlook 2020."
100. PwC (2021). "Global CEO Survey: A year of transformation."
101. UNESCO (2020). "Education and the digital divide: Challenges and opportunities."
102. Ellen MacArthur Foundation. (2021). "Completing the Picture: How the Circular Economy Tackles Climate Change."
103. International Energy Agency (IEA). (2020). "Data Centres and Data Transmission Networks."
104. United Nations. (2019). "Global sustainable development report 2019: The future is now - science for achieving sustainable development."
105. World Economic Forum. (2021). "The Role of Critical Minerals in Clean Energy Transitions." Harvard Business School. (2020). "How the COVID-19 Pandemic is Changing the Way We Work."
106. Global E-waste Monitor (2020). "Quantities, flows, and the circular economy requirements."
107. Capgemini (2019). "The Digital Advantage: How digital leaders outperform their peers in sales growth and profitability."

108. Cybersecurity Ventures. (2021). "Cybercrime to Cost the World $10.5 Trillion Annually by 2025."
109. Deloitte (2021). "The future of work: A journey to digital transformation."
110. Fernanda, C., et al. (2020). "The Impact of Digitalization on SMEs: Findings from the Research."
111. McKinsey (2020). "The COVID-19 recovery will be digital: A plan for the future."
112. McKinsey (2021). "Consumer Decision Journey in a Digital World."
113. PwC. (2020). "AI and automation: How people can outsmart machines."
114. Statista (2022). "Global retail e-commerce sales growth from 2014 to 2026."
115. Capgemini (2018). *The digital transformation review.*
116. IFAD. (2020). *Digital Agriculture in Africa: A Review of the Evidence.*
117. McKinsey & Company (2016). *The digital transformation of industries.*
118. PwC. (2017). *What's driving the growth of digital businesses?.*
119. ITU. (2020). *Measuring digital development: Facts and .*
120. Armbrust, M., et al. (2010). "Above the Clouds: A Berkeley View of Cloud Computing." UC Berkeley.
121. Brynjolfsson, E., & McAfee, A. (2014). *The Second Machine Age: Work, Progress, and Prosperity in a Time of Brilliant Technologies*. W. W. Norton & Company.
122. Castells, M. (1996). *The Rise of the Network Society*. Blackwell Publishing.
123. Gartner (2017). "Top 10 Strategic Technology Trends for 2017."
124. Gartner (2020). "Forecast Analysis: Information Security, Worldwide."
125. Hammer, M., & Champy, J. (1993). *Reengineering the Corporation: A Manifesto for Business Revolution*. HarperBusiness.

126. Hase, S., & Wilson, G. (2003). "The Role of the Informal Learning in the Workplace." *Journal of Workplace Learning*.

127. Kotter, J. P. (1996). *Leading Change*. Harvard Business Review Press.

128. McKinsey & Company (2011). "Big Data: The Next Frontier for Innovation, Competition, and Productivity."

129. Norman, D. A. (2013). *The Design of Everyday Things: Revised and Expanded Edition*. Basic Books.

130. Smit, P. J., et al. (2016). "The Digital Enterprise: Getting Started with Digital Transformation." *Business Horizons*.

131. Armbrust, M., et al. (2010). "Above the Clouds: A Berkeley View of Cloud Computing." UC Berkeley.

132. Brynjolfsson, E., & McAfee, A. (2014). *The Second Machine Age: Work, Progress, and Prosperity in a Time of Brilliant Technologies*. W. W. Norton & Company.

133. Castells, M. (1996). *The Rise of the Network Society*. Blackwell Publishing.

134. Gartner (2017). "Top 10 Strategic Technology Trends for 2017."

135. Gartner (2020). "Forecast Analysis: Information Security, Worldwide."

136. Hammer, M., & Champy, J. (1993). *Reengineering the Corporation: A Manifesto for Business Revolution*. HarperBusiness.

137. Hase, S., & Wilson, G. (2003). "The Role of the Informal Learning in the Workplace." *Journal of Workplace Learning*.

138. Kotter, J. P. (1996). *Leading Change*. Harvard Business Review Press.

139. McKinsey & Company (2011). "Big Data: The Next Frontier for Innovation, Competition, and Productivity."

140. Norman, D. A. (2013). *The Design of Everyday Things: Revised and Expanded Edition*. Basic Books.

Smit, P. J., et al. (2016). "The Digital Enterprise: Getting Started with Digital Transformation." *Business Horizons*.

141. Coursera. (2020). *About MOOCs*.
142. Education Week. (2020). *How Ed Tech is Changing the Classroom*.
143. Harvard Business Review. (2020). *The Future of Artificial Intelligence in Education*.
144. OECD. (2020). *Education 2030: Shaping the Future of Education*.
145. UNESCO. (2020). *ICTs in Education: A Review of the Literature*.
146. Ellen MacArthur Foundation. (2021). "Completing the Picture: How the Circular Economy Tackles Climate Change."
147. International Energy Agency (IEA). (2020). "Data Centres and Data Transmission Networks."
148. United Nations. (2019). "Global sustainable development report 2019: The future is now - science for achieving sustainable development."
149. World Economic Forum. (2021). "The Role of Critical Minerals in Clean Energy Transitions."
150. Harvard Business School (2020). "How the COVID-19 Pandemic is Changing the Way We Work."
151. Global E-waste Monitor (2020). "Quantities, flows, and the circular economy requirements."
152. Capgemini (2019). "The Digital Advantage: How digital leaders outperform their peers in sales growth and profitability."
153. Cybersecurity Ventures. (2021). "Cybercrime to Cost the World $10.5 Trillion Annually by 2025."
154. Deloitte (2021). "The future of work: A journey to digital transformation."
155. Fernanda, C., et al. (2020). "The Impact of Digitalization on SMEs: Findings from the Research."

156. FAO. (2019). *The State of Food and Agriculture 2019: Moving Forward on Food Loss and Waste Reduction.*

157. FAO. (2020). *Proceedings of the Global Conference on Animal Welfare: Opportunities and Challenges.*

158. FAO. (2021). *The State of the World's Land and Water Resources for Food and Agriculture.*

159. IFAD. (2020). *The Role of Livestock in Rural Development: A Review.*

160. Kenya Dairy Board (2021). *Annual Report 2020-2021.*

161. PAM. (2019). *Global Report on Food Crises.*

162. World Organisation for Animal Health (2020). *World Animal Health Overview 2020.*

163. European Commission. (2020). *EU Agricultural Outlook for Markets and Income 2020-2030.*

164. Atomico (2021). State of European Tech 2021. Atomico.

165. Gartner. (2021). Forecast Analysis: Public Cloud Services, Worldwide. Gartner.

166. PitchBook (2021). Private Equity and Venture Capital Trends. PitchBook.

167. KPMG. (2022). The Growing Global Fintech Ecosystem. KPMG.

168. Grand View Research (2021). Health Tech Market Size, Share & Trends Analysis Report. Grand View Research.

169. HolonIQ. (2021). Global EdTech Market Size & Share. HolonIQ.

170. Hired (2021). The State of Tech Hiring. Hired.

171. African Union. (2019). FTAA: An opportunity for Africa. African Union.

172. African Development Bank. (2020). Intraregional trade in Africa: opportunities and challenges. AfDB .

173. World Bank. (2021). Monthly African economic outlook. World Bank.

174. Ministry of Planning and Development, Ethiopia. (2018). Vision 2030: Strategy for sustainable and inclusive development.

175. Ethiopian Investment Commission. (2021). Investment report 2021.

176. World Economic Forum (2021). Ethiopia: challenges and opportunities for 2030. WEF.

177. Rwanda Development Board (2020). National Development Plan: Agenda 2050. RDB.

178. Rwanda Utilities Regulatory Authority (2021). Energy and sustainable development in Rwanda. RURA.

179. Amnesty International. (2021). Rwanda 2021 Human Rights Report. Amnesty.

180. African Union. (2019). FTAA: An opportunity for Africa. African Union.

181. African Development Bank. (2020). Intraregional trade in Africa: opportunities and challenges. AfDB .

182. World Bank. (2021). Monthly African economic outlook. World Bank.

183. Ministry of Planning and Development, Ethiopia. (2018). Vision 2030: Strategy for sustainable and inclusive development.

184. Ethiopian Investment Commission. (2021). Investment report 2021.

185. World Economic Forum (2021). Ethiopia: challenges and opportunities for 2030. WEF.

186. Rwanda Development Board (2020). National Development Plan: Agenda 2050. RDB.

187. Rwanda Utilities Regulatory Authority (2021). Energy and sustainable development in Rwanda. RURA.

188. Amnesty International. (2021). Rwanda 2021 Human Rights Report. Amnesty.

189. International Telecommunication Union (ITU). (2021). Report on measuring the digital economy. ITU.

190. UNESCO. (2020). Report on digital education in Africa. UNESCO.

191. World Bank. (2021). Poverty and inequality in Africa. World Bank.

192. McKinsey & Company (2020). The impact of digitization on employment in Africa. McKinsey.

193. UNICEF. (2021). Education, pandemics and access to technology. UNICEF.

194. Ernst & Young (2020). Innovation and digital skills in Africa. EY.

195. Government of Rwanda (2018). Smart Rwanda: a strategy for the knowledge economy. Government of Rwanda.

196. African Union. (2020). Initiatives for Internet access in Africa. African Union.

197. International Telecommunication Union (ITU). (2021). Measuring Digital Development: Facts and Figures 2021. ITU.

198. World Bank. (2020). Digital Economy for Africa: A Strategy for Transforming Africa's Economy. World Bank.

199. McKinsey & Company (2019). The Economic Impact of Digital Infrastructure. McKinsey.

200. GSMA. (2020). The Mobile Economy 2020. GSMA.

201. UNESCO. (2020). Education during COVID-19 and beyond. UNESCO.

202. World Bank. (2021). Digital Inclusion: A Pathway to Recovery. World Bank.

203. Shirky, C. (2017). The End of Audience: The Changing Nature of News in the Digital Age. *The New Yorker*.

204. World Economic Forum. (2021). Harnessing the Power of Technology for Sustainable Development. WEF.

205. International Telecommunication Union (ITU). (2021). Measuring Digital Development: Facts and Figures 2021. ITU.

206. World Bank. (2021). Connecting the World: The Role of Energy in Digital Development. World Bank.

207. McKinsey & Company (2020). The Future of Rural Connectivity: How Digital Transformation Can Foster Economic Growth. McKinsey.

208. Cohen, K., & Halperin, M. (2020). Building Bridges: Addressing the Digital Divide in Rural Areas. *Harvard Kennedy School Review*.

209. GSMA. (2020). The Mobile Economy 2020. GSMA.

210. Fischer, U. (2019). Connecting the Unconnected: Options for Rural Infrastructure Development. *International Journal of Development Studies*, 7(2), 150-165.

211. Ministry of Information and Communication Technology, India (2020). Digital India: A Programme to Transform India. Government of India.

212. SpaceX (2021). Starlink: A New Internet from Space. SpaceX.

213. Community Networks (2020). Building Community Networks in South Africa. Community Networks.

214. Organisation for Economic Co-operation and Development (OECD). (2021). Going Digital in a Post-COVID World: Innovations in Connectivity. OECD.

215. World Economic Forum. (2021). The Future of Economic Engagement in Rural Areas: Digitization and Connectivity. WEF.

McBratney, A., Whelan, B., Ancev, T., & Bouma, J. (2005). Future directions of precision agriculture. *Soil Use and Management, 21*(4), 363-370.

216. Gebbers, R., & Adamchuk, V. I. (2010). Precision agriculture and food security. *Science, 327*(5968), 1111-1112.

217. Food and Agriculture Organization (FAO). (2018). The State of Food and Agriculture: Leveraging Food Systems for Inclusive Rural Transformation. Retrieved from FAO Report.

218. Food and Agriculture Organization (FAO). (2019). Digital Agriculture: Technology and Innovations in Agriculture. Retrieved from FAO Report.

219. World Bank. (2016). Analysis of Agriculture in Sub-Saharan Africa: Financial and Institutional Approaches for Improved Productivity. Retrieved from World Bank Report.

220. European Union (EU). (2020). Innovative Technologies and Practices in Sustainable Agriculture: A European Perspective. Retrieved from EU Report.

221. Alexander, C., & McLaren, J. (2015). The Role of GPS in Precision Agriculture. *Journal of Precision Agriculture, 16*(1), 55-70.

222. Anderson, K., & Gaston, S. (2013). Lightweight Drones for Ecological Applications. *Ecological Applications, 23*(5), 1214-1225.

223. Bannari, A., et al. (2018). The Role of Models in Precision Agriculture. *Agricultural Systems, 160*, 37-50.

224. Bongiovanni, R., & Lonardi, C. (2004). Precision Agriculture and Sustainability: An Overview. *Proceedings of the International Conference on Precision Agriculture, 32*, 10-20.

225. Khoshnevisan, B., et al. (2019). Farm Management Information Systems: Current State and Future Directions. *Computers and Electronics in Agriculture, 162*, 388-395.

226. Khan, K. A., et al. (2019). Challenges in the Adoption of Precision Agriculture Technologies: A Comprehensive Review. *Agriculture, 9*(1), Article 9.

227. Anderson, K., & Gaston, S. (2013). Lightweight Drones for Ecological Applications. *Ecological Applications, 23*(5), 1214-1225.

228. Friedrich, T., et al. (2019). Remote Sensing Applications in Precision Agriculture: A Review. *Remote Sensing, 11*(1), 50.

229. Khoshnevisan, B., et al. (2019). Farm Management Information Systems: Current State and Future Directions. *Computers and Electronics in Agriculture, 162*, 388-395.

230. Kumar, A., et al. (2018). Drip Irrigation Management in Agriculture: A Review. *Irrigation Science, 36*(4), 361-377.

231. Schwarz, K. M., et al. (2015). Advances in Irrigation Technology for Precision Agriculture. *Water Management, 27*(3), 125-140.

232. Wang, J., et al. (2017). Soil Sensor Technologies for Precision Agriculture: A Review. *Computers and Electronics in Agriculture, 141*, 259-266.

233. Chase, S., et al. (2014). "Evaluation of UAV-based Photogrammetry for Mapping Vegetation in a Tidal Marsh." *Remote Sensing, 6*(1), 164-182.

234. Colomina, I., & Molina, P. (2014). "Unmanned Aerial Systems for Photogrammetry and Remote Sensing: A Review." *ISPRS Journal of Photogrammetry and Remote Sensing, 92*, 79-97.

235. Crampton, J. W. (2009). "Cartography: Maps 2.0". *Progress in Human Geography, 33*(2), 235-252.

236. Dandois, J. P., & Ellis, E. C. (2010). "Remote Sensing for Biodiversity: The Use of UAVs to Evaluate Plant Species Diversity." *Ecosphere, 1*(4), 1-15.

237. Gonzalez, A., et al. (2020). "The Economic Impact of Using UAV in Agricultural Practices." *European Journal of Remote Sensing, 53*(1), 35-50.

238. Kärcher, M. & Klöckner, C. (2016). "High-Resolution Mapping Using UAVs." *Journal of Geographical Systems, 18*(3), 329-345.

239. Kahn, D. et al. (2013). "Historical Development of Unmanned Aerial Vehicles." *The Journal of Transportation Technology, 7*, 533-548.

240. McKinley, J., et al. (2020). "Drones in Disaster Relief: Technology and Policy Dilemmas." *International Journal of Disaster Risk Reduction, 51*, 101834.

241. Murphy, R. (2020). "The Future of Drone Technology: Trends and Predictions." *Journal of Unmanned Vehicle Systems, 8*(2), 56-62.

242. Robo, A., et al. (2019). "The Use of Drones in Modern Agriculture: A Comprehensive Review." *Agricultural Systems, 172*, 147-157.

243. Zhang, C., & Kovacs, J. M. (2012). "The Application of Small Unmanned Aerial Systems for Precision Agriculture: A Review." *Precision Agriculture, 13*(6), 693-712.

244. Chase, S., et al. (2014). "Evaluation of UAV-based Photogrammetry for Mapping Vegetation in a Tidal Marsh." *Remote Sensing, 6*(1), 164-182.

245. Colomina, I., & Molina, P. (2014). "Unmanned Aerial Systems for Photogrammetry and Remote Sensing: A Review." *ISPRS Journal of Photogrammetry and Remote Sensing, 92*, 79-97.

246. Crampton, J. W. (2009). "Cartography: Maps 2.0". *Progress in Human Geography, 33*(2), 235-252.

247. Dandois, J. P., & Ellis, E. C. (2010). "Remote Sensing for Biodiversity: The Use of UAVs to Evaluate Plant Species Diversity." *Ecosphere, 1*(4), 1-15.

248. Gonzalez, A., et al. (2020). "The Economic Impact of Using UAV in Agricultural Practices." *European Journal of Remote Sensing, 53*(1), 35-50.

249. Kärcher, M. & Klöckner, C. (2016). "High-Resolution Mapping Using UAVs." *Journal of Geographical Systems, 18*(3), 329-345.

250. Kahn, D. et al. (2013). "Historical Development of Unmanned Aerial Vehicles." *The Journal of Transportation Technology, 7*, 533-548.

251. McKinley, J., et al. (2020). "Drones in Disaster Relief: Technology and Policy Dilemmas." *International Journal of Disaster Risk Reduction, 51*, 101834.

252. Murphy, R. (2020). "The Future of Drone Technology: Trends and Predictions." *Journal of Unmanned Vehicle Systems, 8*(2), 56-62.

253. Robo, A., et al. (2019). "The Use of Drones in Modern Agriculture: A Comprehensive Review." *Agricultural Systems, 172*, 147-157.

254. Zhang, C., & Kovacs, J. M. (2012). "The Application of Small Unmanned Aerial Systems for Precision Agriculture: A Review." *Precision Agriculture, 13*(6), 693-712.

255. Bashar, M. A., et al. (2020). "IoT-based Health Monitoring System: A Review." *Journal of Ambient Intelligence and Humanized Computing, 12*(6), 5229-5242.

256. Bertino, E., & Islam, N. (2017). "Botnets and Internet of Things Security." *Computer, 50*(2), 76-79.

257. Gubbi, J., et al. (2013). "Internet of Things (IoT): A Vision, Architectural Elements, and Future Directions." *Future Generation Computer Systems, 29*(7), 1645-1660.

258. Kumpulainen, S., et al. (2020). "Smart Home Energy Management System Using IoT." *Journal of Energy Storage, 28*, 101424.

259. Miorandi, D., et al. (2012). "Internet of Things: Vision, Applications and Research Challenges." *Ad Hoc Networks, 10*(7), 1497-1516.

260. Murphy, R. (2020). "The Future of IoT in Smart Cities." *Journal of Urban Technology, 27*(2), 1-16.

261. Sundmaeker, H., et al. (2010). "Vision and Challenges for Realizing the Internet of Things." *European Commission.*

262. Wolfert, S., et al. (2017). "Big Data in Smart Farming - A Review." *Agricultural Systems, 153*, 69-80.

263. Zanella, A., et al. (2014). "Internet of Things for Smart Cities." *IEEE Internet of Things Journal, 1*(1), 22-32.

264. Bramas, M., et al. (2018). "Farm Management Information Systems: A Review of Concepts and Applications." *Agricultural Systems, 162*, 61-76.

265. Klerkx, L., et al. (2019). "The Role of Digital Technologies in the Transformation of Agriculture." *Journal of Agriculture and Environmental Ethics, 32*(5), 767-784.

266. López, L. C., et al. (2019). "Precision Agriculture and Farm Management: Advancements and Future Directions." *Sustainability, 11*(18), 5002.

267. Thompson, D. L., et al. (2018). "The Impact of Precision Agriculture on Farm Operations." *Journal of Management Studies, 55*(4), 721-747.

268. Blue River Technology (2021). "Our Technology: See How We're Redefining Agriculture." [Online]. Available: Blue River Technology

269. California Department of Water Resources. (2020). "California's Water Use Approaches." [Online]. Available: CDWR

270. AgriPredict (2020). "Leveraging AI for Enhanced Agricultural Productivity."

271. FarmCrowdy (2019). "Investing in Food Security: A Report on Our Impact."

272. M-KOPA. (2021). "Transforming Agriculture with Solar Power in Kenya."

273. World Bank. (2021). "Digital Agriculture in Africa: Transforming Agriculture with Technology."

274. Challenges and Prospects." *Journal of Agricultural Science and Technology*, 23(2), 123-137.

275. Kumar, L., Muthusamy, K., & Sharma, R. (2021). "Harnessing Big Data and Artificial Intelligence in Agriculture." *Global Perspectives on Sustainable Agriculture*, 34(4), 645-659.

276. McKinsey & Company (2019). "How AI can Help Agriculture: Connecting the Dots." Report accessed at: mckinsey.com

277. Schmidt, J., Winter, S., & Jensen, M. (2020). "Artificial Intelligence in Agriculture: Addressing the Threats to Food Security and Sustainability." *Agriculture and Human Values*, 37(2), 477-491.

278. Wang, Q., Sun, Y., & Zhang, Y. (2020). "Ethical Issues in the Application of Artificial Intelligence in Agriculture." *AI and Ethics*, 1(2), 179-188.

279. FAO (2020). "Digital Technologies in Agriculture and Rural Areas." Report available at: fao.org

280. Boston Consulting Group (2019). "The Future of Agriculture: How Digital Technologies are Transforming Agriculture."

281. FAO. (2021). "Digital Technologies in Agriculture and Rural Areas." Report available at: fao.org.

282. Kumar, L., Muthusamy, K., & Sharma, R. (2020). "Harnessing Big Data and Artificial Intelligence in Agriculture." *Global Perspectives on Sustainable Agriculture*, 34(4), 645-659.

283. Meyer, E., Fuchs, C., & Hohfeld, L. (2019). "Digital Transformation in Agriculture: Current Adoption Trends and Future Perspectives." *Journal of Agricultural Engineering*, 50(3), 132-142.

284. Poudel, B., Lantagne, S., & Sabourin, E. (2020). "Barriers to Adoption of Digital Technologies in Agriculture: A Case Study of Smallholder Farmers in Nepal." *Technology in Society*, 62, 101297.

285. ITU. (2020). "Measuring Digital Development: Facts and Figures 2020." International Telecommunication Union (ITU) report.

286. Brynjolfsson, E., & McAfee, A. (2014). *The Second Machine Age: Work, Progress, and Prosperity in a Time of Brilliant Technologies*. W. W. Norton & Company.

287. Chui, M., Manyika, J., & Miremadi, M. (2016). "Where machines could replace humans-and where they can't (yet)." *McKinsey Quarterly*.

288. Davenport, T. H., & Ronanki, R. (2018). "AI for the Real World." *Harvard Business Review*, 96(1), 108-116.

289. Kaplan, J., & Haenlein, M. (2019). "Siri, Siri, in my Hand: Who's the Fairest in the Land? On the Interplay of Trust and AI in User-Technology Interaction." *Business Horizons*, 62(3), 375-384.

290. Müller, V. C., & Bostrom, N. (2016). "Future Progress in Artificial Intelligence: A Survey of Expert Opinion." *Global Priorities Project*.

291. World Economic Forum. (2020). "The Future of Jobs Report 2020." Retrieved from weforum.org.

292. Bertolini, M., et al. (2019). "The Role of Public and Private Partnerships in Artificial Intelligence Development." *AI & Society*, 34(1), 109-116.

293. European Commission. (2021). "Proposal for a Regulation on a European Approach for Artificial Intelligence." Brussels.

294. CNIL. (2021). "Règlement Général sur la Protection des Données." Commission Nationale de l'Informatique et des Libertés.

295. Government of Canada (2020). "Artificial Intelligence for Business: Funding Opportunities." Government of Canada.

296. Krause, J., et al. (2020). "Artificial Intelligence in Education: A Global Perspective." *Education and Information Technologies*, 25(2), 479-494.

297. Agricultural Data Coalition. (2020). "Unlocking the Value of Agricultural Data." Agricultural Data Coalition.

298. Davis, K., & He, Y. (2020). "The Role of Digital Agriculture in Supporting the Agriculture Sector." *Agricultural Systems*, 180, 102790.

299. Mavrakis, D., et al. (2020). "Digital Agriculture: A New Era in Agriculture." *International Journal of Information and Computer Security*, 12(1), 51-65.

300. Patel, V., et al. (2019). "Collaboration between Governments and Private Sector: The Path to Innovation in Agriculture." *International Journal of Agricultural Management*, 9(3), 130-140.

301. Sustainable Development Solutions Network.(2020). "Building a Sustainable Future with AI in Agriculture." SDSN.

302. Tey, Y. S., & Brindal, M. (2012). "Factors Influencing the Adoption of Precision Agriculture in Australia." *Precision Agriculture*, 13, 48-71.

303. World Bank. (2019). "Digital Agriculture: Opportunities for Africa." World Bank.

304. World Economic Forum. (2021). "The Future of Agriculture: The Role of AI and Digital Technologies." WEF.

305. Baker, J. (2019). *Water Management in Precision Agriculture: New Technologies for Better Crop Yields*. Journal of Irrigation and Drainage Engineering, 145(3), 04019035.

306. Cook, S., et al. (2020). *Economic Impacts of Precision Agriculture: A Case Study from the Midwest*. Agricultural Economics, 51(2), 233-245.

307. FAO. (2019). *The Future of Food and Agriculture: Trends and Challenges*. Rome: Food and Agriculture Organization.

308. Kammerer, W., & Saracoglu, A. (2021). *Technology and Skills in the Future of Agriculture: Adapting to Change*. International Journal of Agricultural Management, 10(2), 74-86.

309. Lobell, D. B., et al. (2014). *Climate Change and Threats to Food Production*. Science, 346(6206), 1232-1235.

310. Meyer, J., et al. (2020). *Barriers to Adoption of Precision Agriculture Technologies: An Exploration of the Literature*. Precision Agriculture, 21(5), 1230-1245.

311. Zhang, X., et al. (2019). *Precision Agriculture in Crop Management: Enhancing Productivity and Sustainability*. Journal of Agricultural Science, 157(5), 747-758.

312. Bertot, J. C., Jaeger, P. T., & Grimes, J. M. (2016). *Promises and realities of big data in public administration: A case study of the E-Government program*. Journal of E-Government Studies and Best Practices, 2016, 1-11.

313. Buntong, S., et al. (2019). *Mobile market platforms and rural farmer income in Kenya: Randomized controlled trials of M-Farm*. Journal of Agricultural Economics, 70(3), 735-754.

314. FAO. (2020). *The State of Food and Agriculture: Overcoming the Digital Divide in Rural Areas*. Rome: Food and Agriculture Organization.

315. Jack, W., & Suri, T. (2014). *Risk sharing and transaction costs: Evidence from Kenya's mobile money revolution*. American Economic Review, 104(3), 183-207.

316. OECD. (2019). *The Digital Economy: Policies for Growth and Jobs in Rural Areas*. Paris: Organisation for Economic Co-operation and Development.

317. ITU. (2021). *ICT Facts and Figures 2021*. Geneva: International Telecommunication Union.

318. World Bank. (2017). *Digital Dividends: World Development Report 2016*. Washington, DC: World Bank.

319. Agarwal, S., et al. (2021). *Access to Finance in the Agricultural Sector: Opportunities and Challenges*. Agricultural Systems, 190, 103173.

320. FAO. (2019). *The State of Agricultural Commodity Markets: Agricultural Value Chain Development*. Rome: Food and Agriculture Organization.

321. FAO. (2020). *Digital Technologies for Agriculture: Transforming Value Chains for Sustainable Development*. Rome: FAO.

322. Gollin, D., et al. (2016). *Agricultural Productivity Gaps in Developing Countries: A Comparison of 10 Countries and their Agricultural Value Chains*. World Development, 78, 197-207.

323. IPCC. (2021). *Climate Change 2021: The Physical Science Basis. Contribution of Working Group I to the Sixth Assessment Report of the Intergovernmental Panel on Climate Change*. Cambridge University Press.

324. Kassam, A., et al. (2019). *The role of agroecology in strengthening agricultural value chains*. Agronomy for Sustainable Development, 39(3), 1-12.

325. Norton, G. W., et al. (2019). *Agricultural Value Chain Management: A Review of the Literature*. Journal of Agricultural and Resource Economics, 44(1), 149-179.

326. Olsen, S. B., et al. (2020). *Infrastructure Needs for Agricultural Value Chains: A Maryland Case Study*. Journal of Infrastructure Systems, 26(4), 04020041.

327. World Bank. (2019). *The Role of Value Chains in Leveraging Agriculture for Development*. Washington, DC: World Bank.

328. World Bank. (2020). *Creating Markets in Africa: Value Chains in Food and Agriculture*. Washington, DC: World Bank.

329. World Resources Institute. (2020). *Creating a Sustainable Food Future: A Menu of Solutions to Feed Nearly 10 Billion People by 2050*. Washington, DC: WRI.

330. Murphy, D. (2019). *Artificial Intelligence: A Guide to Intelligent Systems.* 3rd Edition. Addison-Wes

331. Brynjolfsson, E., & McAfee, A. (2014). *The Second Machine Age: Work, Progress, and Prosperity in a Time of Brilliant Technologies.* W. W. Norton & Company.

332. Chui, M., Manyika, J., & Mrkvička, T. (2017). *Where machines could replace humans-and where they can't (yet).* McKinsey Global Institute.

333. Luckin, R. (2018). *Artificial Intelligence and Education: The Importance of Teacher and Student Relationships.* Routledge.

334. Susskind, R., & Susskind, D. (2015). *The Future of the Professions: How Technology Will Transform the Work of Human Experts.* Harvard University Press.

335. Kaplan, J. (2016). "Artificial Intelligence: The Revolution Hasn't Happened Yet." *Harvard Data Science Review.*

336. Bessen, J. E. (2019). "AI and Jobs: The Role of Demand." *NBER Working Paper Series.*

337. OECD. (2018). "OECD Digital Economy Outlook 2017."

338. "AI & Machine Learning for Coders." (2023). Google Developers

339. "Artificial Intelligence and the Future of Work." (2023). McKinsey & Company

340. Gonzalez, A. (2021). "Digital Transformation & Artificial Intelligence in the Workplace." Doctoral dissertation, Stanford University.

341. TED Talks (series on AI and digitalization). TED

342. Kamilaris, A., & Prenafeta-Boldú, F. X. (2018). *Deep Learning in Agriculture: A Survey*. Computers and Electronics in Agriculture.

343. Liakos, K. G., et al. (2018). *Agricultural robotics and the use of artificial intelligence in agriculture: A review*. Computers and Electronics in Agriculture.

344. Nassour, A., & Barnett, J. (2020). *Sustainable agriculture: Integrating precision agriculture and smart technologies*. Sustainable Agriculture Reviews.

345. Zhang, C., et al. (2019). *Digital agriculture: Transforming agriculture through digital technologies*. Journal of Agricultural and Food Information.

346. Capgemini (2019). "The Digital Advantage: How digital leaders outperform their peers in sales growth and profitability."

347. Cybersecurity Ventures. (2021). "Cybercrime to Cost the World $10.5 Trillion Annually by 2025."

348. Deloitte (2021). "The future of work: A journey to digital transformation."

349. Fernanda, C., et al. (2020). "The Impact of Digitalization on SMEs: Findings from the Research."

350. McKinsey (2020). "The COVID-19 recovery will be digital: A plan for the future."

351. McKinsey (2021). "Consumer Decision Journey in a Digital World."

352. PwC. (2020). "AI and automation: How people can outsmart machines."

353. Statista (2022). "Global retail e-commerce sales growth from 2014 to 2026."

354. Agence Nationale de la Statistique et de la Démographie (ANSD). Recensement Général de la Population et de l'Habitat, de l'Agriculture et de l'Elevage. Rapport définitif. 2013.

355. United Nations Conference on Trade and Development. Africa Economic Development Report: Unleashing the potential of Africa's trade in services for trade and development. 2015.

356. B. Diagana and A. Mankor. "Sustainable Agriculture and Poverty Reduction in the Groundnut Basin of Senegal: Results of the Tradeoffs Analysis Model". 2008.

357. International Fund for Agricultural Development. Creating opportunities in the rural non-farm economy. In: Rural Poverty Report 2011, chapter 6. Rome: IFAD. 2011.

358. Y. Haami and Y. Godo. The Three Agricultural Problems in the Disequilibrium of World Agriculture", Journal of Agriculture and Development, 3-16. 2004.

359. United Nations. Recent social trends in Africa. In "Economic Report on Africa. 2017.

360. World Bank. Growing the Rural Non-Farm Economy to Alleviate Poverty: An IEG Evaluation of World Bank Group Support 2004-2014," Approach Paper. December 8, 2015.

LIST OF FIGURES

Figure 1: Intelligent irrigation system 6

Figure 2: How Agritech and digital culture are transforming agriculture in Uganda 11

Figure 3: Plant disease verification using the IA application 82

Figure 4: Using androids to list in the database database 83

Figure 5: Precision watering 84

Figure 6: Precision farming concept 89

Figure 7: Agtech or Agritech 90

Figure 8: Precision agriculture organization chart 93

Figure 9: Producing more with less: An agricultural model reserved for the wealthy 94

Figure 10: Using drones for reduced inspections 101

Figure 11: Referencing system 102

Figure 12: Promising IoT sensors for emerging agriculture 109

Figure 13: Crop accuracy and geo-referencing 118

Figure 14: The three (3) stages of precision 123

Figure 15: Crop disease detection....................... 157

Table of contents

PREFACE v

ABBREVIATIONS vii

Summary viii

Introduction 1

Chapter 1: Digitization in Africa and its impact on rural economies 10

1-1-Definition and key concepts 13

1-1-1-Digitization: definition and components 13

1-1-2- Rural economy in Africa: characteristics and challenges. 32

1-2-The state of digitalization in Africa 52

1-2-1- Internet and digital technology penetration rate 69

1-2-2- Digital infrastructure in rural areas 73

1-3-Impact of digitalization on rural economies 80

1-3-1-Improving market access 81

1-3-2- Reducing transaction costs 82

1-3-3-Strengthening agricultural value chains 82

Chapter 2: Precision agriculture in Africa 89

2-1- Introduction to precision farming 91

2-1-1-Definition and components of precision agriculture 92

2-1-2-Key technologies used 94

2-2- Overview of precision agriculture in Africa 123

2-2-1- Adoption of precision farming technologies 129

2-2-2- Factors facilitating and constraining adoption 134

2-3-Impact of precision farming on crop yields 137

2-3-1- Case study in Africa 143

2-3-2-Comparative analysis of yields before and after adoption 147

Chapter 3: The role of artificial intelligence in precision agriculture 150

3-1-1-Presentation of artificial intelligence and its applications in agriculture 153

3-1-2- Concrete examples of AI use in precision agriculture 156

3-2- Impact of AI on the efficiency of precision agriculture 158

3-2-1-Analysis of productivity gains 160

3-2-2-Reducing operating costs and optimizing resources 162

3-3-Case study on the impact of AI on precision agriculture in Africa 164

3-3-1-Practical use of AI in agricultural projects in Africa 164

3-3-2-Results and lessons learned 166

Chapter 4: Challenges and prospects for AI integration in agriculture 168

4-1-Challenges of AI integration 169

4-1-1-Digital infrastructure problem 172

4-1-2-Cost and accessibility of technologies 175

4-1-3- Lack of technical expertise 178

4-2-Perspective and recommendations 182

4-2-1- Public policies to encourage AI adoption 183

4-2-2-Reinforcing local capacity 185

4-2-3-Public-private partnership for the development of agricultural techniques 186

Conclusion 190

BIBLIOGRAPHY..205

Printed by Books on Demand GmbH, Norderstedt / Germany